9th European Summer School on Experimental Nuclear Astrophysics 2017

EPJ Web of Conferences Volume 184 (2018)

Catania, Italy
17 – 24 September 2017

Editors:

C. Spitaleri
L. Lamia
R.G. Pizzone
G. Rapisarda
M.L. Sergi

ISBN: 978-1-5108-6595-2

Printed from e-media with permission by:

Curran Associates, Inc.
57 Morehouse Lane
Red Hook, NY 12571

Some format issues inherent in the e-media version may also appear in this print version.

Printed by Curran Associates, Inc. (2018)

For additional information, please contact EDP Sciences – Web of Conferences at the address below.

EDP Sciences – Web of Conferences
17, Avenue du Hoggar
Parc d'Activité de Courtabœuf
BP 112
F-91944 Les Ulis Cedex A
France

Phone: +33 (0) 1 69 18 75 75
Fax: +33 (0) 1 69 28 84 91

contact-edps@webofconferences.org

Additional copies of this publication are available from:

Curran Associates, Inc.
57 Morehouse Lane
Red Hook, NY 12571 USA
Phone: 845-758-0400
Fax: 845-758-2633
Email: curran@proceedings.com
Web: www.proceedings.com

TABLE OF CONTENTS

R-PROCESS OBSERVATIONS 1
Aoki Wako
COSMOLOGICAL LITHIUM PROBLEMS 7
Bertulani C.A., Chintak Shubh, Mukhamedzhanov A.M.
THE LUNA EXPERIMENT: PAST AND FUTURE 16
Broggini Carlo
THE S-PROCESS IN STELLAR SITES 22
Cristallo Sergio
THE α-DECAY OF THE HOYLE STATE IN ^{12}C: A NEW HIGH-PRECISION INVESTIGATION 28
Dell'Aquila D., Lombardo I., Verde G., Vigilante M., Acosta L., Agodi C., Cappuzzello F., Carbone D., Cavallaro M., Cherubini S., Cvetinovic A., D'Agata G., Francalanza L., Guardo G.L., Gulino M., Indelicato I., La Cognata M., Lamia L., Ordine A., Pizzone R.G., Puglia S.M.R., Rapisarda G.G., Romano S., Santagati G., Spartà R., Spadaccini G., Spitaleri C., Tumino A.
RESONANT ELASTIC SCATTERING 36
de Oliveira Santos Francois
HEAVY ELEMENT NUCLEOSYNTHESIS 46
El Eid Mounib F.
TROJAN HORSE METHOD EXPERIMENTS WITH RADIOACTIVE ION BEAMS 53
Gulino Marisa, Cherubini Silvio, Gabriele Rapisarda Giuseppe, Kubono Shigeru, Lamia Livio, La Cognata Marco, Gianluca Pizzone Rosario, Yamaguchi Hidetoshi, Hayakawa Seya, Wakabayashi Yasuo, Iwasa Naohito, Kato Seigo, Komatsubara Tetsuro, Teranishi Takashi, Coc Alain, de Séréville Nicolas, Hammache Fairouz, Kiss Gabor, Bishop Shawn, Nguyen Binh Dam, Roeder Brian, Trache Livius, Tribble Robert, Spartà Roberta, Indelicato Iolanda, Spitaleri Claudio
TRANSFER REACTIONS FOR NUCLEAR ASTROPHYSICS 59
Hammache Fairouz
EXPERIMENTAL APPROACH TO EXPLOSIVE HYDROGEN BURNING IN X-RAY BURSTS AND CORE-COLLAPSE SUPERNOVAE 73
Kubono Shigeru
PRIMORDIAL NUCLEOSYNTHESIS: CONSTRAINTS ON THE BIRTH OF THE UNIVERSE 79
Mathews Grant, Kusakabe Motohiko, Gangopadhyay Mayukh, Kajino Toshitaka, Sasankan Nishanth
RIB PRODUCTION AND RELATED EXPERIMENTS AT EXOTIC 89
Mazzocco Marco
ESSENTIALS OF THE MACROSCOPIC-MICROSCOPIC FOLDED-YUKAWA APPROACH AND EXAMPLES OF ITS RECORD IN PROVIDING NUCLEAR-STRUCTURE DATA FOR SIMULATIONS 94
Möller Peter
ANC EXPERIMENTS FOR NUCLEAR ASTROPHYSICS IN NPI CAS 102
Mrazek J., Burjan V., Kroha V., Mukhamedzhanov A.M., Tribble R., Spitaleri C., Sivácek I., Glagolev V., Piskor Š., Pizzone G.R., La Cognata M., D'Agata G., Tumino A., Lamia L., Sparta R., Rapisarda G., Romano S.
GAS DETECTORS FOR NUCLEAR PHYSICS EXPERIMENTS 108
Pierroutsakou Dimitra
THE TROJAN HORSE METHOD IN NUCLEAR ASTROPHYSICS 114
Tumino Aurora, Spitaleri Claudio, Cherubini Silvio, D'Agata Giuseppe, Giovanni Luca Guardo, Gulino Marisa, Indelicato Iolanda, La Cognata Marco, Lamia Livio, Rosario Gianluca Pizzone, Giuseppe Gabriele Rapisarda, Romano Stefano, Maria Letizia Sergi, Spartá Roberta
DIRECT MEASUREMENTS AND DETECTION TECHNIQUES WITH LOW-ENERGY RIBS 120
Yamaguchi H., Hayakawa S., Yang L., Shimizu H., Kahl D.
PROTON-INDUCED REACTIONS OF ASTROPHYSICAL INTEREST 126
Chillery Thomas
IMPROVED INFORMATION ON ASTROPHYSICAL S-FACTOR FOR THE ^{10}B(P,α_0)^{7}BE REACTION USING THE TROJAN HORSE METHOD 130
Cvetinovic A., Spitaleri C., Spartá R., Rapisarda G.G., Puglia S.M.R., La Cognata M., Cherubini S., Guardo G.L., Gulino M., Lamia L., Pizzone R.G., Romano S., Sergi M.L., Tumino A.

THE ^{19}F(α, P)^{22}NE AND ^{23}NA(P, α)^{20}NE REACTION IN AGB NUCLEOSYNTHESIS VIA THM 134
D'Agata G., Pizzone R. G., La Cognata M., Indelicato I., Spitaleri C., Burjan V., Cherubini S., Di Pietro A., Guardo G. L., Gulino M., La Commara M., Lamia L., Lattuada M., Mazzocco M., Mrazek J., Milin M., Palmerini S., Parascandolo C., Pierroutsakou D., Rapisarda G.G., Romano S., Sergi M.L., Soic N., Spartá R., Trippella O., Tumino A.

^{7}BE(N,P) CROSS SECTION MEASUREMENT FOR THE COSMOLOGICAL LITHIUM PROBLEM AT THE N_TOF FACILITY AT CERN 139
Lucia Anna Damone, Colonna N., Barbagallo M., Mastromarco M., Andrzejewski J., Finocchiaro P., Cosentino L.

FIRST MEASUREMENT OF ^{72}GE(N, γ) AT N_TOF 144
Dietz M., Lederer-Woods C., Aberle O., Andrzejewski J., Audouin L., Bacak M., Balibrea J., Barbagallo M., Becvár F., Berthoumieux E., Billowes J., Bosnar D., Brown A., Caamaño M., Calviño F., Calviani M., Cano-Ott D., Cardella R., Casanovas A., Cerutti F., Chen Y. H., Chiaveri E., Colonna N., Cortés G., Cortés-Giraldo M. A, Cosentino L., Damone L. A., Diakaki M., Domingo-Pardo C., Dressler R., Dupont E., Durán I., Fernández-Dominguez B., Ferrari A., Ferreira P., Finocchiaro P., Furman V., Göbel K., García A. R., Garg R., Gawlik A., Gilardoni S., Glodariu T., Goncalves I. F., González-Romero E., Griesmayer E., Guerrero C., Gunsing F., Harada H., Heinitz S., Heyse J., Jenkins D. G., Jericha E., Käppeler F., Kadi Y., Kahl D., Kalamara A., Kavrigin P., Kimura A., Kivel N., Kokkoris M., Krticka M., Kurtulgil D., Leal-Cidoncha E., Leeb H., Lerendegui-Marco J., Lo Meo S., Lonsdale S. J, Macina D., Marganiec J., Martínez T., Masi A., Massimi C., Mastinu P., Mastromarco M., Maugeri E. A, Mazzone A., Mendoza E., Mengoni A., Milazzo P. M, Mingrone F., Musumarra A., Negret A., Nolte R., Oprea A., Patronis N., Pavlik A., Perkowski J., Porras I., Praena J., Quesada J. M, Radeck D., Rauscher T., Reifarth R., Rubbia C., Ryan J. A, Sabateé-Gilarte M., Saxena A., Schillebeeckx P., Schumann D., Sedyshev P., Smith A. G, Sosnin N. V, Stamatopoulos A., Tagliente G., Tain J. L, Tarifeño-Saldivia A., Tassan-Got L., Valenta S., Vannini G., Variale V., Vaz P., Ventura A., Vlachoudis V., Vlastou R., Wallner A., Warren S., Weiss C., Woods P. J, Wright T., Žugec P.

DEVELOPMENT OF THE ELISSA ARRAY: PROTOTYPE TESTING AT LABORATORI NAZIONALI DEL SUD 149
Guardo G. L., Anzalone A., Balabanski D., Chesnevskaya S., Crucillá W., Filipescu D., Gulino M., La Cognata M., Lattuada D., Matei C., Pizzone R. G., Rapisarda G. G., Romano S., Spitaleri C., Taffara A., Tumino A., Xu Y.

STUDY OF THE CONTRIBUTION OF THE ^{7}BE(D, P) REACTION TO THE ^{7}LI PROBLEM IN THE BIG-BANG NUCLEOSYNTHESIS 152
Inoue A., Tamii A., Abe K., Adachi S., Aoi N., Asai M., Fukuda M., Gey G., Hashimoto T., Ideguchi E., Isaak J., Kobayashi N., Maeda Y., Makii H., Matsuta K., Mihara M., Miura M., Shima T., Shimizu H., Tang R., Dinh Trong T., Yamaguchi H., Yang L.

A GEANT4-BASED MONTE CARLO TOOL FOR NUCLEAR ASTROPHYSICS 156
Lattuada D., La Cognata M., Anzalone A., Balabanski D.L., Chesnevskaya S., Costa M., Crucillà V., Guardo G.L., Gulino M., Matei C., Pizzone R.G., Romano S., Spitaleri C., Tumino A., Xu Y

BIG BANG NUCLEOSYNTHESIS (BBN) AND NON-STANDARD PHYSICS 159
Makki Tahani, El Eid Mounib

STUDY OF KEY RESONANCES IN THE ^{30}P(P,γ)^{31}S REACTION IN CLASSICAL NOVAE 162
Meyer A., de Séréville N., Hammache F., Adsley P., Assié M., Beaumel D., Delafosse C., Flavigny F., Georgiadou A., Gottardo A., Grassi L., Guillot J., Id Barkach T., MacCormick M., Matea I., Olivier L., Perrot L., Portail C., Stefan I., Parikh A., Coc A., Kiener J., Tatischeff V., Laird A. M., Fox S. P., Hubbard N., Riley J., De Oliveira F., Bastin B., Béroff K., Sánchez Benítez Á. M., Alellara A., Assunção M., Guimaraes V., Oulebsir N., D'Agata G.

RESONANCE EFFECTS IN CARBON BURNING PROCESS ON TYPE IA SUPERNOVAE 166
Mori Kanji, Famiano Michael, Kajino Toshitaka

THE TREIMAN-YANG CRITERION: VALIDATING THE TROJAN HORSE METHOD BY EXPERIMENTALLY PROBING THE REACTION MECHANISM 169
Perrotta S. S., Spitaleri C., Cherubini S., Cvetinovic A., D'Agata G., Dell'Aquila D., Di Pietro A., Figuera P., Guardo L., Gulino M., Indelicato I., Kres I., La Cognata M., La Commara M., Lamia L., Lattuada D., Lattuada M., Lombardo I., Mazzocco M., Parascandolo T., Pizzone R. G., Rapisarda G. G., Romano S., Spartà R., Trippella O., Tumino A.

ISOMERIC ^{26}AL BEAM PRODUCTION WITH CRIB 173
Shimizu H., Kahl D., Yamaguchi H., Abe K., Beliuskina O., Cha S. M., Chae K. Y., Chen A. A., Ge Z., Hayakawa S., Imai N., Iwasa N., Kim A., Kim D. H., Kim M. J., Kubono S., Kawag M. S., Liang J., Moon J. Y., Nishimura S., Oka S., Park S. Y., Psaltis A., Teranishi T., Ueno Y., Yang L.

^{26}MG TARGET FOR NUCLEAR ASTROPHYSICS MEASUREMENTS 176
Sivácek I., Mrázek J., Kroha V., Burjan V., Glagolev V., Piskor Š., Spitaleri C., Pizzone R.G., La Cognata M., D'Agata G., Tumino A., Lamia L., Spartá R., Rapisard G.G., Romano S.

^{7}BE AND ^{8}B REACTION DYNAMICS AT COULOMB BARRIER ENERGIES 180
Strano E., Mazzocco M., Boiano A., Boiano C., La Commara M., Manea C., Parascandolo C., Pierroutsakou D., Signorini C., Torresi D., Yamaguchi H., Kahl D., Acosta L., Di Meo P., Fernandez-Garcia J.P., Glodariu T., Grebosz J., Guglielmetti A., Imai N., Hirayama Y., Ishiyama H., Iwasa N., Jeong S.C., Jia H.M., Keeley N., Kim Y.H., Kimura S., Kubono S., Lay J.A., Lin C.J., Marquinez-Duran G., Marte I., Miyatake H., Mukai M., Nakao T., Nicoletto M., Pakou A., Rusek K., Sakaguchi Y., Sanchez-Benitez A.M., Sava T., Sgouros O., Stefanini C., Soramel F., Soukeras V., Stiliaris E., Stroe L., Teranishi T., Toniolo N., Wakabayashi Y., Watanabe Y.X., Yang L., Yang Y.Y.

Author Index

In this book a collection of the lecture notes given during the Nineth European Summer School on Experimental Nuclear Astrophysics is given. The school belongs to the European Network of Nuclear Astrophysics Schools (ENNAS), a network made by the European schools on nuclear astrophysics and related areas (the Santa Tecla, the Sinaia, the Russbach schools), having the common effort of preparing and educating young physicists in nuclear physics, astrophysics and their mutual relationship in the nuclear astrophysics field. The school, whose first edition was held in 2001, took place from 17 to 24 of September 2017 in Santa Tecla, a small village about 15 km north of Catania, characterized by its position on the volcanic shores of the Ionian Sea, surrounded by the spectacular Timpa area, a green protected park specific for its mediterranean vegetation. 70 between young students and researchers from more than 20 countries of all the world attended the lectures and were also encouraged to present their work work and results.

The school has tried once more to present to the young students the global picture of nuclear astrophysics research in the last years. Thus the scientific program of the school covered a wide range of topics dealing with various aspects of nuclear astrophysics, such as stellar evolution and nucleosynthesis, neutrino physics, Big Bang, direct and indirect methods, radioactive ion beams. Nuclear astrophysics plays a key role for understanding energy production in stars, stellar evolution and the concurrent synthesis of the chemical elements and their isotopes. It is also a fundamental tool to explain the ashes of the early universe, to determine the age of the universe through the study of pristine stellar objects and to predict the evolution for the Sun or Stars. The bone structure for the above aspects is based on nuclear reactions, whose rates need to be determined in laboratories. Although impressive progress has been made over the past decades, which was rewarded by Nobel prizes, several open questions are still unsolved, which challenge the basis of the present understanding.

A list of the lecture topics is given below:

- Big Bang Nucleosynthesis;
- Plasmas in stars and laboratoties;
- Radioactive ion beams;
- Detector and facilities for nuclear astrophysics;
- Indirect methods in nuclear astrophysics;

An often quoted statement of Carl Sagan, "We are all star stuff", is exhaustively describing one of the great discoveries of the XX century as well as the main aim of nuclear astrophysics. The present theory of stellar nucleosynthesis (based on the famous paper B2FH) predicts chemical evolution of the Universe, which is testable by looking at stellar spectral lines, meteorites, pre- solar grains or other means of investigation. Quantum mechanics explains why different atoms emit light at characteristic wavelengths, and so by studying the light emitted from different stars, one may infer the atmospheric composition of individual stars.

Big bang nucleosynthesis tells us that the early universe consisted of only the light elements, and so one expects the first stars to be composed of hydrogen, helium, and lithium, the three lightest elements. The recent achievements of WMAP and Planck missions, with the precise measurement of many cosmological parameters,have reignited the interest on the primordial nucleosynthesis, especially for the still unclear Lithium primordial abundance. Stellar structure and the H-R diagram indicate that the lifetime of a star depends greatly on its initial mass and chemical composition, so that massive stars are very short-lived, and less massive stars are longer-lived. As a star dies, nuclear astrophysics argues that it will enrich the interstellar medium with heavy elements (in this case all elements heavier than lithium, the third element), from which new stars are formed. This account is consistent with the observed correlation between stellar metallicity and red shift.

The theory of stellar nucleosynthesis would not be very convincing if reliable nuclear physics inputs are adopted. By carefully scrutinizing the table of nuclides, nuclear astrophysicists were able to predict the existence of different stellar environments which could produce the observed isotopic abundances, and the nuclear processes which must occur in these stars. This is also valid for elements heavier than iron production, when fusion reactions are of negligible importance and the main nucleosynthetic path goes through p-process, r-process, and s-process. Many of the nuclides generated in the huge explosion triggered by the last burnings in the life of a massive star are unstable. Then in order to understand what is going on in Supernovae explosions, Gamma-ray bursts, Novae it was pointed out that the physics of radioactive beams should be explored. Many group of physicists have then taken the opportunities offered by the world-wide developments of radioactive ion beams facilities to start to investigate such environments. In this field also the weak interaction processes are important and in some cases, like type II Supernovae, dominant. This is extremely important in the multimessenger astronomy era.

In order to better understand our Universe, a precise knowledge of nuclear inputs for nuclear astrophysics is therefore needed. They influence sensitively the nucleosynthesis of the elements in the early stages of the universe and in all the objects formed thereafter and control the associate energy generation, neutrino luminosity and stellar evolution. Thus a good knowledge of reaction rates is essential to understand this broad picture. Studies at the energies typical of nuclear astrophysics for charged particle induced reactions (few keV - 1 MeV) are severely hampered by the low signal-to-noise ratio, essentially due to the tiny cross sections that have to be measured (as low as picobarn in some cases). The best experimental solutions to this main problem will be discussed in this book and presented in their details. The pilot project, LUNA, i.e. a 50 keV accelerator installed in the Laboratori Nazionali del Gran Sasso has already given several important results in the last decade. Other experimental methods such as the recoil separator, spectrometers will be reviewed together with the wide field of nuclear astrophysics research opened by the development of radioactive ion beams in several laboratories of the world.

Detectors and facilities useful for the next decade of nuclear astrophysics studies were reviewed in the lectures and a big attention was devoted to future research possibilities for studies. This knowledge and facilities will be of groundbreaking importance for solving the future quests for this field. Although all these efforts in some cases (such as the $^{12}C+^{12}C$ interaction, the $^{12}C+^{16}O$ and many others) the goal is far from being achieved by means of direct measurements. Moreover the presence of the electron screening effect makes very hard and sometimes impossible the measurement of the bare nucleus astrophysical factor, that is the quantity needed for astrophysics. Thus extrapolations (usually supported by R-matrix theory) are adopted.

An alternative way to by-pass these problems and measure the bare nucleus astrophysical factor or cross section is given by the indirect methods. Among them the Coulomb Dissociations, the ANC, decay spectroscopy and the Trojan Horse Method will be reviewed. In recent years, new possibilities of exploring the universe have opened. Not only electromagnetic spectra, luminosities are available for stars, galaxies and compact objects but new windows are opened in particle detection. Cosmic rays detectors will play in the near future and important role for a better understanding of what is going on in the universe and therefore they will trigger the opening of a new sector of nuclear astrophysics research. Physics experiments can be performed in laser-induced plasmas available worldwide, opening new possibilities for future research.

Crucial problems related to various issues of nuclear as well as astrophysics are still open, triggering new researches in this field and the enthusiasm to organize events such us our European Summer school on experimental nuclear astrophysics. Once more students, young researchers and lecturers have gathered in Santa Tecla from more than twenty countries of the world. The school, though an European one, keeps playing the role of a meeting point of students of different nationalities from the whole world. These students, we are sure, have left our school not only with an improved knowledge of nuclear astrophysics but also with new friends (as well as future collaborators) all around the world. We gratefully acknowledge all the participants the lecturers and the members of the Scientific committee.The school could not have taken place without their efforts and help. The Organizing Committee also acknowledges the financial support of the Laboratori Nazionali del Sud - Istituto Nazionale di Fisica Nucleare - Italy, Dipartimento di Fisica e Astronomia of the University of Catania - Italy and ENSAR2. The school has been honoured by the welcome addresses of the school director prof. Claudio Spitaleri, by Dr. Giacomo Cuttone, Director of Laboratori Nazionali del Sud, by prof. S. Romano on behalf of Dipartimento di Fisica e Astronomia, by prof. F. Priolo, president of the Scuola Superiore di Catania and by prof. A. Insolia, Director of INFN-Sezione Catania. We wish to the reader that the reading of this book can be as fruitful as the participation to our school.

C. Spitaleri
L. Lamia
R.G. Pizzone
G.G. Rapisarda
M.L. Sergi

Statement of Peer review

In submitting conference proceedings to *Web of Conferences*, the editors of the proceedings certify to the Publisher that

1. They adhere to its **Policy on Publishing Integrity** in order to safeguard good scientific practice in publishing.
2. All articles have been subjected to peer review administered by the proceedings editors.
3. Reviews have been conducted by expert referees, who have been requested to provide unbiased and constructive comments aimed, whenever possible, at improving the work.
4. Proceedings editors have taken all reasonable steps to ensure the quality of the materials they publish and their decision to accept or reject a paper for publication has been based only on the merits of the work and the relevance to the journal.

Title, date and place of the conference

9th European Summer School on Experimental Nuclear Astrophysics
Santa Tecla, 17-24 September 2017

Proceedings editor(s):

Spitaleri Claudio
Lamia Livio
Pizzone Rosario Gianluca
Rapisarda Giuseppe
Sergi Maria Letizia

Date and editor's signature

31.05.2018

r-process observations

Wako Aoki[1,⋆]

[1] *National Astronomical Observatory of Japan, 2-21-1 Osawa, Mitaka, Tokyo 181-8588, Japan*

Abstract. The r-process nucleosynthesis in the Universe is constrained by observations of chemical abundances of the Solar System and those of old stars that should record the products of the r-process events in the early Universe. This review provides a brief overview of the observational technique to determine chemical abundances of stars. Targets of observations are stars in different populations of the Milky Way Galaxy and surrounding dwarf galaxies, providing different kind of constraints on the understanding of the r-process. Recent progress in observational studies to identify the r-process sites are also reviewed.

1 Introduction

The r-process is a major source of heavy neutron-capture elements in the universe. This process should be related to rare explosive events with extreme conditions, like neutron star mergers or some sort of supernova explosions. A goal of current observational studies on the r-process is to identify the astrophysical site of this process, or at least to provide useful constraints on its characteristics.

Measurements of heavy neutron-capture elements in explosive events by optical spectroscopy to measure spectral lines, or by observations of γ-ray from radio-active nuclei, are desired to identify the sites of the r-process and to determine the production yields. There is, however, no definitive detection of heavy neutron-capture elements enhanced in promising r-process sites. Measurements of heavy elements in absorption spectra of supernova remnants were attempted, resulting in no clear detection [1]. Very recently, a merger of binary neutron stars was identified both by gravitational wave and electromagnetic waves including γ-ray and optical light [2][3]. The excess of the red and near-infrared flux in the observed afterglow is suggested to be a result of ejection of lantanides, which are much more opaque in the optical range than usual metal like Fe [4]. Detection of some neutron-capture elements in optical spectra of the afterglow is also suggested, though it contains large uncertainties, and further measurements in the near future and modeling of the events are desirable.

On the other hand, observational constraints on the r-process currently available have been indirectly obtained from abundance measurements for solar-system material and stars in the Milky Way and nearby galaxies.

2 Constraints from solar-system abundances

The elemental abundances of solar-system material have been used as the most important reference to compare the prediction of r-process models (e.g., [5, 6]).

⋆e-mail: aoki.wako@nao.ac.jp

The elemental abundances of solar-system material are determined from chemical analysis of meteorites and analysis of the solar spectrum. Recent measurements by the two methods exhibit fairly good agreement for heavy neutron-capture elements in general [7]. In particular, the recent updates of atomic line data, including the effect of isotope shift and hyperfine splitting, have been making large contribution to the improvements of solar atmospheric abundances (e.g., [8]).

An advantage of the analysis of meteorites is that it can determine isotope ratios, which provide much stronger constraints on the models of neutron-capture processes than elemental abundance ratios. It should be noted that isotope ratios of a few heavy elements (e.g. Eu) have been determined by the detailed profile analysis of spectral lines which are affected by hyperfine splitting depending on isotopes (e.g. [9, 10]).

The heavy neutron-capture elements with mass number (A) larger than ~ 70 in the solar-system material are originated from both the s- and r-processes. The contributions of the two processes are well identified by the existence of abundance peaks corresponding to the neutron magic numbers (50, 82, and 126), appearing slightly different mass numbers due to the difference of the reaction paths on nuclear chart between the two processes.

Since the abundances of s-process yields are basically determined by reactions along stable nuclei, modeling of this process is better established than that of the r-process. There are dozens of nuclei which are never produced by the r-process because they are shielded by other stable nuclei from β-decay of neutron-rich unstable nuclei ("s-only" nuclei). Models of the s-process are calibrated by the abundance of these s-only nuclei, which enable us to estimate the fraction of abundances produced by the s-process for other nuclei [11]. The r-process component of each isotope is obtained by subtracting the s-process component from the abundance of each isotope for solar-system material.

Hence, the uncertainty of the s-process models affects the accuracy of the estimates of the r-process abundance patterns of solar-system material. This is significant in the nuclei in which s-process component is dominant, because a small error in the estimate of the s-process component could results in a large error in the obtained r-process fraction. An example is Pb, for which the s-process component is estimated to be as large as 80% in solar-system material. This makes it difficult to constrain the models for production of heaviest stable nuclei by the r-process [12].

3 Abundance patterns of metal-poor stars

The solar abundances discussed in the previous section are determined by numerous nucleosynthesis processes in the long history of the Milky Way before the formation of the Sun at around 4.6 billion years ago. Hence, the r-process abundance pattern, which is separated from the s-process component, should be regarded as a sum of the r-process yields by a large number of events.

By contrast, measurements of heavy neutron-capture elements in very metal-poor stars provide unique constraint on the r-process models. Such stars are regarded as old low-mass stars formed in the early stage of the Milky Way evolution, at which the metallicity of gas clouds was generally very low. Metal-poor stars formed from such clouds should record nucleosynthesis yields of only a small number of events. In the most extreme cases, the abundance of neutron-capture elements could be a result of a single event. The abundance patterns of such objects are very useful references to study the yields of the r-process and their possible variation.

A remarkable result obtained by the intensive spectroscopic studies for very metal-poor stars in the past two decades is the discovery of large star-to-star scatter in abundance ratios of neutron-capture elements. Figure 1 shows the Mg and Eu abundances as a function of metallicity (Fe abundance)[1]. Here we adopt Eu as an indicator of the r-process yields.

[1] Abundance ratios of two elements are given as $[A/B] = \log(N_A/N_B) - \log(N_A/N_B)_\odot$. Here "Fe abundance" means [Fe/H].

We find a clear correlation with little scatter in abundances between Mg and Fe, both of which are yields of massive stars and their supernova explosions in the low metallicity range. Contributions of type Ia supernovae are significant at high metallicity, that result in a change of the slope in [Fe/H]> −1 (see next section). By contrast, the Eu abundances show large star-to-star scatter at low metallicity. Metal-poor stars having high Eu abundances are interpreted as objects that were born from gas clouds significantly polluted by an r-process event in the early Galaxy. This indicates that the gas clouds were not chemically homogeneous due to insufficient mixing in the early Galaxy, and the r-process events should be independent of those yielding Fe, i.e. usual core-collapse supernovae.

The detection probability of Eu is lower in metal poor stars having low Eu abundances, because Eu spectral lines are quite weak in such objects. Given this bias in the sample, Eu-enhanced stars are rare objects. This indicates that the r-process events are rare, but each event could provide the Universe with a large amount of Eu.

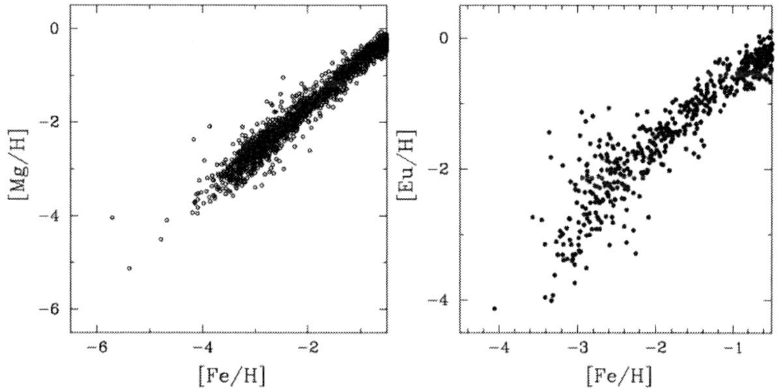

Figure 1. Abundance ratios of Mg and Eu with respect to Fe. Abundance data are taken from the SAGA database [13].

Eu-enhanced metal-poor stars are ideal objects to study the abundance patterns produced by the r-process. A difficulty in the spectroscopic measurement of neutron-capture elements is the weakness of spectral lines of these elements and severe contamination of features of other elements (e.g., Fe, Ti) and molecules. This difficulty is reduced in the analysis of metal-poor objects having excess of r-process elements including Eu. As a result, many spectral lines of neutron-capture elements that are not measured even for the solar spectrum are used in the analysis of Eu-enhanced metal-poor stars.

A remarkable result obtained for Eu-rich stars is that abundance patterns of neutron-capture elements of these stars are very similar to that of the r-process component in solar-system material [14, 15]. The agreement is in particular evident in elements from the second and third peaks corresponding to the neutron magic numbers 82 and 126 (Ba-Pt). We note that the measurements of the elements at the second r-process peak (Te-Xe) is very limited because there is no useful spectral lines in the optical range [16].

This result indicates that the r-process produces neutron-capture elements with similar abundance patterns in every event. This phenomenon is called the universality of the r-process, and has a large impact on understanding of the process, because it could be a strong constraint on r-process models that predict more or less variation of abundance ratios depending on model parameters like electron fraction and entropy.

It should be noted that the abundance patterns of light neutron-capture elements (e.g., Sr, Y) are also similar between such Eu-enhanced stars, but they show some deviation from the solar r-process pattern [15]. Moreover, there are many metal-poor stars that have low abundances of heavy neutron-capture elements like Eu, but have large excess of light ones [17]. A recent study [18] has revealed that metal-poor stars have wide variations in the abundance patterns from light to heavy neutron-capture elements including intermediate ones (e.g., Mo, Pd). This suggests that there are at least two origins of the r-process component of light neutron-capture elements in the universe, and both were effective even in the early Galaxy.

4 Abundance trend and scatter

The trends and scatter in the Eu abundance distribution also provide a useful constraint on the origins of the r-process. The enrichment of an element depends on the timescale of the corresponding nucleosynthesis event. For instance, as found in Figure 1, Mg abundance ratios are systematically higher than Fe abundance ratios by 0.3–0.5 dex in the low-metallicity range. This reflects the yields from massive stars and their supernova explosions. On the other hand, a large fraction of Fe in the solar-system are originated from type Ia supernovae, whose time-scale is longer than core-collapse supernovae of massive stars.

Although the scatter of Eu abundance ratios in metal-poor stars is large, the existence of Eu-enhanced objects at low metallicity suggests that the time-scale of the r-process is not long. This was regarded as a difficulty in the scenario of merger of binary neutron stars, for which timescale of about 100 million years is expected [19].

Recent chemical evolution models ([20, 21]), however, predict that neutron-capture elements could be enhanced by merger of binary neutron stars even at low metallicity, including accretion of small stellar systems originated from mini-halos in the formation process of the Milky Way halo. In such small systems, star formation, and as a result metal enrichment, should have proceeded more slowly than in large systems. Hence, it takes relatively long time in formation of very metal-poor stars, some of which could have suffered from the r-process events by neutron star mergers.

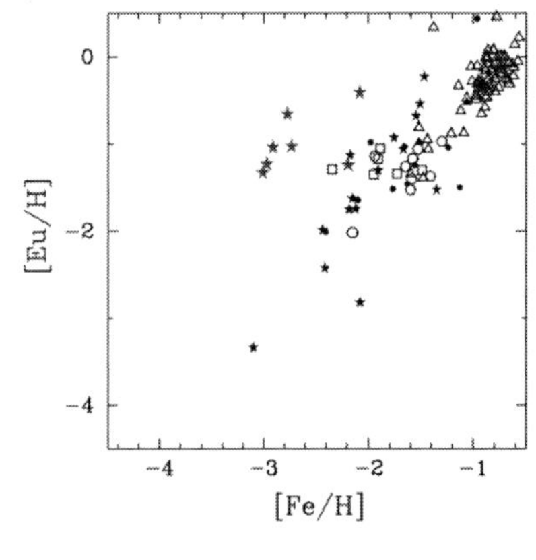

Figure 2. Eu abundance ratios of stars in dwarf galaxies as a function of metallicity ([Fe/H]). Abundance data are taken from the SAGA database [13]. Different symbols mean stars in different galaxies (open triangle: Fornax; open circle: Carina; filled circle: Sculptor; open square: Draco; filled triangle: Leo I; smaller stars: Ursa Minor; large red stars: Reticulum 2). A typical error of [Eu/H] is 0.2 dex, although it depends on data quality and strength of spectral lines. The seven objects in Reticulum 2 are very metal-poor stars that show large excess of Eu.

Small stellar systems are found as satellite dwarf galaxies in the present Milky Way Galaxy, which have wide range of stellar mass (10^3–10^7 solar masses). They could be survivors of small stellar systems formed in the very early phase of the Milky Way formation. Such small stellar systems could be significantly affected by a small number of nucleosynthesis events.

Figure 2 compiles the Eu abundances measured for dwarf galaxy stars. As in the case of field halo stars, the Eu abundances increase with the increase of Fe abundances. The number of stars with low Eu abundances at low metallicity ([Fe/H]< -2) is small, compared to field halo stars in Figure 1, but this should be because of the detection limit of Eu lines in dwarf galaxy stars (see Section 3).

Tsujimoto & Shigeyama [22] pointed out that dwarf galaxy stars with $-2 \lesssim$ [Fe/H] $\lesssim -1$ have almost constant [Eu/H] values, independently of metallicity. Subsequent observational studies [23] for dwarf galaxies also support this feature. The r-process event should have supplied a large amount of neutron-capture elements, but should be a quite rare event compared to supernova explosions that supply Fe and other metals. In larger stellar systems, as in the bulk of the Milky Way halo, more than one events of r-process should have occurred, as well as supernova explosions, resulting in a correlation between Eu and Fe abundances. Neutron star mergers could be the rare events corresponding to the r-process.

Among the faintest dwarf galaxies, called Ultra Faint dwarf Galaxies, Reticulum 2 is a unique object, in which very metal-poor stars ([Fe/H]< -2.5) show large excess of Eu ([Eu/Fe]$> +1$) [24][25]. Such Eu-enhanced stars are known in the Milky Way halo as mentioned above, and their fraction is estimated to be 3–5% (e.g., [26]). The studies for Reticulum 2 measured elemental abundances for nine stars and found seven of them have high Eu abundances. The remaining two stars have the lowest metallicity ([Fe/H]< -3). This suggests that the gas clouds in the whole galaxy was polluted by a single event of the r-process in the early phase before forming very metal-poor stars with [Fe/H]~ -3. Accretion of such stellar systems could be the origins of the r-process enhanced very metal-poor stars in the Milky Way halo.

The Eu abundances found in faint dwarf galaxies, including Reticulum 2, are [Eu/H]~ -1. This corresponds to the Eu/H mass ratio of $10^{-10.3}$. Assuming the original mass of hydrogen gas in the galaxy to be 10^6 solar masses, the Eu mass provided by the r-process is estimated to be $10^{-4.3}$ solar masses. This is comparable to the Eu mass expected to be produced by a single event of merger of binary neutron stars and a magnetrotationally driven supernovae ($10^{-4.5}$–10^{-5} [27]). Assuming the contribution of r-process is only once in individual galaxies, the event rate is 1000-2000 times lower than usual supernovae, as 1000-2000 supernova explosions are expected in a galaxy with a stellar mass of 10^6 solar masses [24]. This rate is also consistent with that expected for neutron star mergers.

5 Summary and concluding remarks

Recent observational studies for very metal-poor stars in the Milky Way halo and dwarf galaxies have been providing useful constraints on the origins and nature of the r-process. Further observations of a large number of stars in individual dwarf galaxies are desirable for better understanding of the astrophysical sites of the r-process and its role in chemical evolution. Such observations covering fainter objects in dwarf galaxies will become possible with new generation large telescopes like TMT, E-ELT and GMT with high-resolution spectrometers. More direct evidence of the r-process will be obtained by future observations of mergers of binary neutron stars following the first detection of the gravitational wave and corresponding electromagnetic waves for GW 170817.

References

[1] G. Wallerstein, A. D. Vanture, E. B. Jenkins, and G. M. Fuller, Astrophys. J. 449, 688 (1995)

[2] A. Goldstein, P. Veres, E. Burns, et al., Astrophys. J., 848, L14 (2017)
[3] E. Pian, P. D'Avanzo, S. Benetti, et al., Nature, 551, 67 (2017)
[4] M. Tanaka, and K. Hotokezaka, Astrophys. J. 775, 113 (2013)
[5] S. Wanajo, & Y. Ishimaru, Nuclear Physics A, 777, 676 (2006)
[6] S. Goriely, A. Bauswein, & H.-T. Janka, Astrophys. J. Letters, 738, L32 (2011)
[7] M. Asplund, N. Grevesse, A. J. Sauval, & P. Scott, Ann. Rev. Astron. Astrophys., 47, 481 (2009)
[8] J. E. Lawler, C. Sneden, J. J. Cowan et al., Astrophys. J. Suppl., 178, 71-88 (2008)
[9] C. Sneden, J. J. Cowan, J. E. Lawler, et al., Astrophys. J., 566, L25 (2002)
[10] W. Aoki, S. Honda, T. C. Beers, & C. Sneden, Astrophys. J., 586, 506 (2003)
[11] F. Kppeler, H. Beer, & K. Wisshak, Reports on Progress in Physics, 52, 945 (1989)
[12] C. Arlandini, F. Käppeler, K. Wisshak, et al., Astrophys. J., 525, 886 (1999)
[13] T. Suda, J. Hidaka, W. Aoki, et al., Publ. Astron. Soc. Japan, 69, 76 (2017)
[14] C. Sneden, A. McWilliam, G. W. Preston, et al., Astrophys. J. 467, 819 (1996)
[15] C. Sneden, J. J. Cowan, and R. Gallino, Ann. Rev. Astron. Ast rophys. 46, 241 (2008)
[16] I. U. Roederer, J. E. Lawler, J. J. Cowan et al., Astrophys. J., 747, L8 (2012)
[17] S. Honda, W. Aoki, Y. Ishimaru, S. Wanajo, & S. G. Ryan, Astrophys. J., 643, 1180 (2006)
[18] M. Aoki, Y. Ishimaru, W. Aoki, & S. Wanajo, Astrophys. J., 837, 8 (2017)
[19] D. Argast, M. Samland, F.-K. Thielemann, and Y.-Z. Qian, Astr on. Astrophys., 416, 997 (2004)
[20] Y. Hirai, Y. Ishimaru, T. R. Saitoh, et al., Astrophys. J., 814, 41 (2015)
[21] Y. Ishimaru, S. Wanajo, & N. Prantzos, Astrophys. J., 804, L35 (2015)
[22] T. Tsujimoto, and T. Shigeyama, Astron. Astrophys. 565, L 5 (2014)
[23] T. Tsujimoto, M. N. Ishigaki, T. Shigeyama, and W. Aoki, Pub. Astron. Soc. Jpn, 67, L3 (2015)
[24] A. P. Ji, A. Frebel, A. Chiti, and J. D. Simon, Nature, 531, 610 (2016)
[25] I. U. Roederer, M. Mateo, J. I. Bailey, et al., Astron. J., 151, 82 (2016)
[26] P. S. Barklem, N. Christlieb, T. C. Beers, et al., Astron. Astrophys. 439, 129 (2005)
[27] N. Nishimura, T. Takiwaki, and F.-K. Thielemann, Astrophys. J., 810, 109 (2015)

Cosmological Lithium Problems

C.A. Bertulani[1,*], *Shubhchintak*[1,**], and *A.M.* Mukhamedzhanov[2,***]

[1]*Department of Physics and Astronomy, Texas A&M University-Commerce, Commerce, TX 75429, USA*
[2]*Department of Physics and Astronomy, Texas A&M University, College Station, TX 77843, USA*

Abstract. We briefly describe the cosmological lithium problems followed by a summary of our recent theoretical work on the magnitude of the effects of electron screening, the possible existence of dark matter parallel universes and the use of non-extensive (Tsallis) statistics during big bang nucleosynthesis. Solutions within nuclear physics are also discussed and recent measurements of cross-sections based on indirect experimental techniques are summarized.

1 Introduction

The cosmological lithium problem has become one of the most intriguing open questions in cosmology due to inconsistencies between observation and calculations based on the standard Big Bang nucleosynthesis (BBN) for the primordial elemental abundances. The BBN model contains a few parameters such as the baryon-to-photon ratio $\eta = n_b/n_\gamma$, the neutron decay time τ_n, and the number of neutrino families N_ν (see, for instance, Ref. [1]). The parameter η relates to the baryon density of the universe by means of $\Omega_0 h^2 \simeq (\eta/10^{-10})/273$, with the Hubble dimensionless parameter h defined through the relation $H_0 = 100h$ km/s/Mpc, the index '0' meaning present time. The anisotropies of the cosmic microwave radiation (CMB) independently determine the value of η [2, 3] when the universe was about 0.3 Myr after the Big Bang. Then photons decoupled and began steaming freely in the universe. Precise LEP experiments to deduce the number of neutrino families [4] lead to the value $N_\nu = 2.9840 \pm 0.0082$, and neutron lifetime measurements have inferred that $\tau_n \simeq 880.2 \pm 1.0$ s [5].

The observed abundances of light elements probe the universe at the very early stages, i.e., 3-20 minutes, of its existence. During this epoch, the light elements D, ^{3}He, ^{4}He, and ^{7}Li were produced and their abundances in selected astrophysical environments are telltales of the BBN epoch. The BBN model predictions also depend on the nuclear reaction network and magnitude of the nuclear cross sections. A few minutes ($\sim$ 3 min) after the Big Bang, deuterons were formed by neutron capture on protons, by means of the reaction p(n,γ)d. The formation of deuterons is strongly dependent on the value of η. Deuterons are promptly destroyed once they are formed leading to the formation of ^{3}He nuclei by means of the d(p,γ)^{3}He and d(d,n)^{3}He reactions. Deuterons also synthesize tritium by means of the d(d,p)t reaction. ^{4}He are then created by the ^{3}He(d,p)^{4}He and t(d,n)^{4}He reactions. In the end, the BBN model predicts that the universe should be composed of about 75% of hydrogen and

*e-mail: carlos.bertulani@tamuc.edu
**e-mail: khajuria1986@gmail.com
***e-mail: akram@comp.tamu.edu

25% of helium with tiny traces of D, ^{3}He, ^{7}Li and ^{6}Li. The foundations of these results rely on the big bang prediction of the neutron-to-proton ratio n/p = 1/7 when the nucleosynthesis started, i.e., the BBN occurred in a proton-rich environment.

Table 1. Nuclear reactions of importance for big bang nucleosynthesis.

$n \leftrightarrow p$	p(n,γ)d	d(p,γ)^{3}He	d(d, p)t
d(d, n)^{3}He	^{3}He(n, p)t	t(d, n)^{4}He	^{3}He(d, p)^{4}He
^{3}He(α,γ)^{7}Be	t(α,γ)^{7}Li	^{7}Be(n, p)^{7}Li	^{7}Li(p,α)^{4}He

The standard BBN model predicts the ^{7}Li/H abundance ratio of the order of 10^{-10} and the ^{6}Li/H abundance ratio of the order of 10^{-14}. Much after the BBN epoch, ^{6}Li and ^{7}Li can be produced in spallation processes by cosmic rays and ^{7}Li can be synthesized in novae or during AGB stars pulsations. In Ref. [6] it was reported that the ^{7}Li abundance is independent of the metallicity in metal-poor stars with small Fe/H abundances relative to the sun. Such stars are warm ($5700 \leq T \leq 6250$ K) metal-poor dwarf stars observed in the galaxy halo. For low metallicity stars the ^{7}Li abundance is nearly constant and this behavior is known as the "Spite plateau" [6]. ^{7}Li is destroyed in red giants with core temperatures in excess of 10^6 K via the reaction ^{7}Li(p, α)^{4}He and that is why white dwarfs at moderate temperatures have been used in such observations. The Spite plateau provides a reasonable evidence that lithium is neither created nor destroyed in warm dwarfs and that such stars display the abundances of primordial ^{7}Li. On the other hand, it is worthwhile mention that recent observations in low-metallicity stars seem to contradict the conclusions drawn from the Spite plateau [7, 8]. The currently accepted value for the ^{7}Li BBN model abundance, calculated using $\eta = (6.07 \pm 0.07) \times 10^{-10}$ [3], corresponds to Li/H = $(4.16 - 5.34) \times 10^{-10}$ [10] while the observations from metal-poor halo stars yields Li/H = $(1.58 + 0.35 - 0.28) \times 10^{-10}$ [8, 9]. This is approximately a factor of 3 lower than expected and is the source of the lithium puzzle.

The second lithium puzzle involves the abundance of ^{6}Li produced during the BBN by means of the ^{2}H(α,γ)^{6}Li reaction. ^{6}Li nuclei formed in stars disappear quickly by means of other reactions. ^{6}Li is also created in cosmic ray interactions, and could also exist in the atmosphere of metal-poor warm dwarfs in the halo of the galaxy, surviving destruction by cosmic rays. But such assumptions are controversial, because they can also apply to ^{7}Li nuclei. The second lithium puzzle relates to the BBN predictions of the isotopic ratio ^{6}Li/^{7}Li $\sim 10^{-5}$ [10, 12], while observations report ^{6}Li/^{7}Li $\sim 5 \times 10^{-2}$ [14]. This puzzle is less robust because of the complexities involved in 3-dimension calculations involving convection and non-local thermodynamical equilibrium, in particular in the photo-sphere of metal-poor stars. Because they might have a large influence on the ^{6}Li/^{7}Li isotopic ratio, such complexities weaken the arguments for the existence of the second lithium problem which in some scenarios yields a better agreement with BBN predictions [15].

2 Nuclear reaction cross sections

During the BBN, the most relevant nuclear reactions are listed in Table 1. This network of reactions resulted in the production of D, ^{3}H, ^{3}He, ^{4}He, ^{6}Li, ^{7}Li and ^{7}Be. Only very small traces of carbon, nitrogen and oxygen were produced at the 10^{-15} – 10^{-25} abundance level. Therefore, there is no

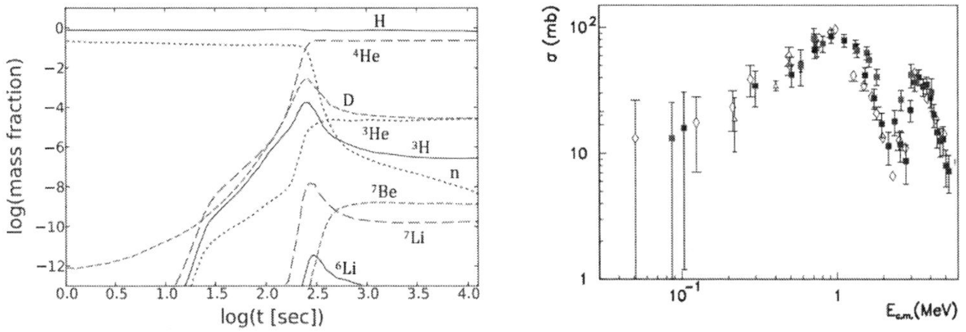

Figure 1. *Left:* Calculated BBN abundances of H, D, ^{3}H, ^{3}He, ^{4}He, ^{6}Li, ^{7}Li and ^{7}Be as a function of time [1]. *Right:* ^{7}Be(n,α) cross section deduced using the THM experimental data for the ^{7}Li(p,α)^{4}He mirror reaction (full red circles) and using ^{3}He THM breakup data (full black circles) [29]. The data compiled by Hou [26] is shown as empty blue circles, and data from Kawabata [28] are shown as full blue squares.

need to include reaction networks beyond those shown in Table 1, such as the famous CNO cycle, to tackle the lithium problem [9]. In Figure 1 we show the calculated BBN abundance of H, D, ^{3}H, ^{3}He, ^{4}He, ^{6}Li, ^{7}Li and ^{7}Be as a function of time [1]. Our calculations were performed with an extended code based on the Wagoner code [16] and similar to NUC123 [17]. The dashed blue curve represents the ^{4}He mass fraction, the red dashed curve represents the deuterium abundance, the green dashed curve represents the ^{3}He abundance, the solid black curve is the ^{3}H abundance, the red dashed curve is the ^{7}Be abundance and blue dashed curve is the ^{7}Li abundance. Recently, new experimental measurements of reactions of relevance for the BBN have been reported based on the use of the Trojan Horse Method (THM) [10].

The reaction cross sections at the low astrophysical energies are enhanced due to the electrons in the plasma. The cross sections in the plasma are enhanced by a factor $f(E) = \sigma_s(E)/\sigma_b(E)$, where σ_s is the screened and σ_b the bare (non-screened) cross section. The Debye-Hückel theory predicts a screened Coulomb potential of the form $V(r) = (e^2 Z_i/r)\exp(-r/R_D)$, where the Debye radius is given by $R_D = (1/\zeta)\left(kT/4\pi e^2 n\right)^{1/2}$, with n being the ion number density and $\zeta = \left[\sum_i X_i(Z_i^2/A_i) + \chi \sum_i X_i(Z_i/A_i)\right]^{1/2}$, with X_i the mass fraction of particle i and the temperature T_6 in units of 10^6 K. χ is a factor correcting for electron degeneracy effects [19]. During the big bang, the electron number density decreased strongly with the temperature, being up to 10^4 times larger than the number density in the core of the sun, $n_e^{sun} \sim 10^{26}/\text{cm}^3$. However, the baryon density was much smaller during the BBN epoch than at the core of the sun. The number of excess electrons during the BBN is nearly the same as those of protons. But most electrons were in balance with the number of positrons produced via $\gamma\gamma \rightarrow e^+e^-$ processes. In Ref. [20] the electron screening effects were included in the BBN reaction network. It was found that the modification of the BBN abundances are negligible. Evidently, it cannot be responsible for the lithium abundance deficiency. Worth mentioning is that recently it has been shown that clustering effects in reactions involving light nuclei at astrophysically relevant energies might also play an important role and could perhaps explain some of the discrepancies found in the experimentally deduced values of electron screening enhancement and theoretical calculations [21].

In Table 2 the BBN calculations are compared with observations. The mass fraction for ^{4}He, historically denoted by Y_p, is taken from Ref. [18], (b) the deuterium abundance D/H = $(2.527 \pm 0.03) \times 10^{-5}$ [22, 23], compatible with $100\Omega_b h^2$ (BBN) = 2.225 ± 0.016 inferred from the measurements of the cosmic microwave background [3], (c) the ^{3}He abundance is taken from Ref. [24], and (d) the lithium abundance is taken from Ref. [8]. The BBN model result for the ^{7}Li abundance shown in Table 2 is in evident discordance (roughly by a factor 3) with the observation. One possibility for this discrepancy could be that ^{7}Be is further destroyed during the BBN. We recall that ^{7}Be decays in 53.22 ± 0.06 days by electron capture to ground and first excited (0.477 MeV) states of ^{7}Li. Therefore, all ^{7}Be produced during the big bang will count towards the ^{7}Li primordial abundance. If ^{7}Be is substantially destroyed by, e.g., (n,p) or (n,α) reactions, it could possibly explain the observed lithium depletion. This possibility has been investigated in Refs. [25–29]. In particular, Lamia et al. [29] have experimentally determined the ^{7}Be(n,α) reaction cross section using the THM experimental data for the ^{7}Li(p,α)^{4}He mirror reaction with corrections for Coulomb effects (see Figure 1). The new deduced data for ^{7}Be(n,α) using this technique lies within the Gamow window appropriate for BBN temperatures and the reaction rate using the new data is found to be lower by a factor ≈ 10 relative to the one used by Wagoner [16]. The new reaction rate yields a ^{7}Li/H abundance ratio of 2.845×10^{-11} and a ^{7}Be/H abundance ratio of 4.156×10^{-10}, leading to a total cosmological lithium abundance of 4.441×10^{-10}, and no appreciable change of the previously obtained BBN results is verified. More recently, a theoretical investigation of the impact of the ^{7}Be(α,γ) on ^{7}Be destruction was performed [30]. It was found that the ^{7}Be abundance would be compromised only if an unexpected strong resonance exists very close to threshold in this reaction channel. All odds are that such a resonant state does not exist.

Table 2. BBN calculations using fits to recent experimental data for BBN reactions compared with observations. The mass fraction for ^{4}He, historically denoted by Y_p, is taken from Ref. [18], (b) deuterium abundance D/H = $(2.527 \pm 0.03) \times 10^{-5}$ [22, 23], compatible with $100\Omega_b h^2$ (BBN) = 2.225 ± 0.016 inferred from the measurements of the cosmic microwave background [3], (c) ^{3}He abundance is taken from Ref. [24], and (d) the lithium abundance is taken from Ref. [8].

Yields	Calculation	Observation
Y_p	0.2485+0.001-0.002	$0.2565 \pm 0.006^{(a)}$
D/H ($\times 10^{-5}$)	2.692+0.177-0.070	$2.527 \pm 0.03^{(b)}$
^{3}He/H ($\times 10^{-6}$)	9.441+0.511-0.466	$\geq 11. \pm 2.^{(c)}$
^{7}Li/H ($\times 10^{-10}$)	4.283 +0.335-0.292	1.58+0.35 -0.28$^{(d)}$

BBN predicts an isotopic ratio of ^{6}Li/^{7}Li $\sim 10^{-5}$, whereas observation yields ^{6}Li/^{7}Li $\sim 5 \times 10^{-2}$ [14]. In Ref. [12] a re-analysis of the reaction ^{4}He(d,γ)^{6}Li was performed, including new predictions for the gamma-ray angular distribution. This was done using a two-body potential model to calculate the S-factor for this reaction at the BBN energies [11, 12]. The potential parameters were chosen to reproduce experimental phase shifts and recently measured ANCs. A nice agreement was found with the experimental data of the LUNA collaboration [13]. This work reinforces BBN predictions for the lithium isotopic ratio and yields a new value of ^{6}Li/^{7}Li = $(1.5\pm0.3)\times 10^{-5}$. The second lithium puzzle seems to be alive although it is not impossible that lithium abundances might change appreciably due to astration.

We conclude this section by stating that it does not seem possible that both lithium puzzles can be solved by accurate measurements of nuclear reaction cross sections, combined with progresses in the theories for nuclear astrophysical reactions. There has been a considerable number of recent theoretical efforts to elucidate the lithium puzzle using a plethora of different ideas based on the premise that physics as we know today might have been different 13.8 billions years ago. New particles, new interactions, changes in fundamental constants, non-standard BBN models, and various intriguing ideas have been used and published elsewhere.

3 Dark matter

Most of the matter in the universe consists of an obscure kind of Dark Matter (DM) which interacts very weakly with the visible matter. In fact, we only know that it interacts gravitationally and large scale experimental searches are underway to identify if DM interacts with visible matter by other means [31–33]. The existence of DM is based on astronomical observations of galaxy clusters dynamics and on the anisotropies of the Cosmic Microwave Background (CMB). Perhaps Weakly Interacting Massive Particles (WIMPs), supersymmetric particles, sterile neutrinos, or any other hitherto undiscovered particles are responsible for its composition. It has also been hypothesized that DM is a mirror sector of particles such as dark photons, dark electrons, etc., which interact in nearly the same way as Standard Model (SM) particles, but only within their own sector. They interact very weakly across sectors, i.e. between the DM sector and the visible sector [34–39, 41]. Besides, the particle copies in the dark sector do not need to have the same masses and couplings as in the visible sector, opening a huge number of possible scenarios for DM.

Astronomical observations yield the ratio of density parameters $\Omega_{DM}/\Omega_{visible} = 4.94 \pm 0.66$. Therefore, DM is 5 times more frequent than visible matter. In Ref. [39–41] this feature was used to explore the possible existence of 5 dark sectors instead of the single ubiquitous dark sector. A Weakly Interacting Massive Gauge Boson (WIMG) was also proposed to couple all dark sectors and ordinary matter. The massive, $E \sim 10$ TeV, WIMG does not modify the properties of the SM and gravity. It has to be consistent with BBN predictions and CMB observations, except maybe with the lithium abundance. Much below the electroweak scale energy, we can assume particles to be massless and group them in matter/charge fields with a similar structure for DM. The WIMG mass is generated by a real scalar field, with the condition that the WIMG has a short-range interaction. In this formalism, the number of dark sectors plays an important role which has been overseen in other BBN models. The new degrees of freedom of particles in the dark sector modify the early universe expansion rate [42] and the elemental abundance predictions. Additional dark sectors increase the effective number of degrees of freedom and their implications for BBN [39–41].

The basic idea of having additional dark sectors is that the radiation in the BBN epoch have densities and entropies given by $\rho(T) = (\pi^2/30)\, g_*(T)\, T^4$ and $s(T) = (2\pi^2/45)\, g_s(T)\, T^3$ with

$$g_*(T) = \sum_B g_B \left(\frac{T_B}{T}\right)^4 + \frac{7}{8}\sum_F g_F \left(\frac{T_F}{T}\right)^4, \quad \text{and} \quad g_s(T) = \sum_B g_B \left(\frac{T_B}{T}\right)^3 + \frac{7}{8}\sum_F g_F \left(\frac{T_F}{T}\right)^3, \qquad (1)$$

where g_* and g_s are the number of degrees of freedom, with $g_{B(F)}$ being the fractions contributed by bosons (fermions) at temperatures $T_{B(F)}$. In this notation, T is the temperature of the radiation thermal bath.

For simplicity, we assume only two temperatures: T in the ordinary matter sector and T' in the dark sectors. By analogy, the energy $\rho'(T')$ and entropy $s'(T')$ densities in the dark sectors are also obtained with Eqs. (1) with $g_*(T) \to g'_*(T')$, $g_s(T) \to g'_s(T')$, and $T \to T'$. An independent variable $x = (s'/s)^{1/3} \sim T'/T$ emerges if one assumes conservation of entropy in all sectors. If each dark sector has the same matter content as in the visible sector, then $g_s(T_0) = g'_s(T'_0)$, leading to $x = T'/T$. The Friedman equation is $H(t) = \sqrt{(8\pi/3c^2)\, G_N\, \bar{\rho}}$, where $\bar{\rho}$ is the total energy density. Including the number of dark sectors, N_{DM}, it becomes $\bar{\rho} = \rho + N_{DM}\, \rho'$. Therefore, one has $H(t) = 1.66\, \sqrt{\bar{g}_*(T)} T^2/M_{Pl}$, with $\bar{g}_*(T) = g_*(T)\left(1 + N_{DM}\, a\, x^4\right)$, where M_{Pl} is the Planck mass and $a = (g'_*/g_*)\,(g_s/g'_s)^{4/3} \sim 1$, for

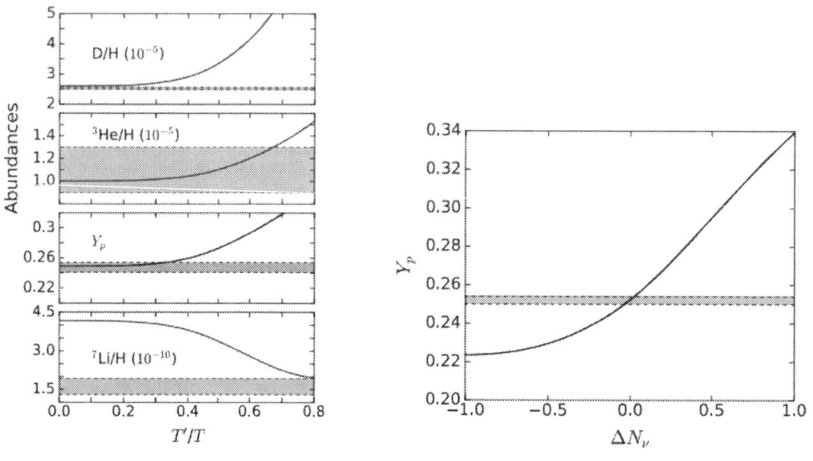

Figure 2. *Left:* Calculated relative abundances of D, ^{3}He, ^{4}He (mass fraction, Y_p) and ^{7}Li as a function of T'/T, with $N_{DM} = 5$ compared to observations. The bands represent uncertainties in the observations [44]. *Right:* Predictions for the primordial ^{4}He mass fraction as a function of extra neutrino families, with $T' = 0.3T_{BBN}$ and $N_{DM} = 5$. The horizontal band represents the observed mass fraction [44].

a not too small T'/T [42]. At about 1 MeV, standard BBN assumes $g_*(T = 1 \text{ MeV}) = 10.75$, but with the additional dark particles it becomes $\bar{g}_* = g_*\left(1 + N_{DM}\, x^4\right)$. We can study the bounds for N_{DM} and x, or T'/T, by comparing BBN calculations and the relative abundances of the light element isotopes (D, ^{3}He, ^{4}He, and ^{7}Li). This is shown in Figure 2 as a function of T'/T with a fixed number of dark sectors, $N_{DM} = 5$. The shaded bands include the uncertainty in the observed values. In this case, we notice that observations of primordial elements of D, ^{3}He, and ^{4}He are compatible with $T'/T \sim 0.2 - 0.3$.

The ^{7}Li problem remains because if $T'/T \sim 1$ then ^{7}Li comes out right, but the other abundances will be completely off the observations. $\bar{g}_*(T)$ is much more sensitive to T' than it is to N_{DM}. Using $T' = 0.3T$ for cold dark sectors, we obtain a large range of values for $N_{DM} = 1 - 50$ compatible with the D, ^{3}He and ^{4}He abundances [44]. Figure 2 also shows the predictions for the primordial ^{4}He mass fraction as a function of extra neutrino families, ΔN_ν, with $T' = 0.3T_{BBN}$ and $N_{DM} = 5$. The horizontal band represents the observed mass fraction [44]. The model is thus compatible with the number of neutrino families $N_\nu = 3$. We thus conclude that there is no incompatibility with the observed primordial abundances and a universe composed with more than one sector of dark matter, e.g. $N_{DM} = 5$ and temperatures of the dark sectors of the order of $T' \sim 0.2 - 0.3T$.

4 Non-extensive statistics

The Maxwell-Boltzmann (MB) distribution is widely known to reproduce extremely well the distribution of velocities of particles in a thermal bath. The MB distribution is a result of the Boltzmann-Gibbs statistics, based on the assumptions that (a) the time between collisions among particles is much larger than their interaction time, (b) the interaction is short-ranged, (c) no correlation exists between the particle velocities, and (d) the collision energy is conserved without transfer to internal degrees of freedom. These very constraining assumptions are not expected to be always valid in thermodynamical equilibrium. In fact, alternatives to the Boltzmann-Gibbs (BG) statistics are known to exist

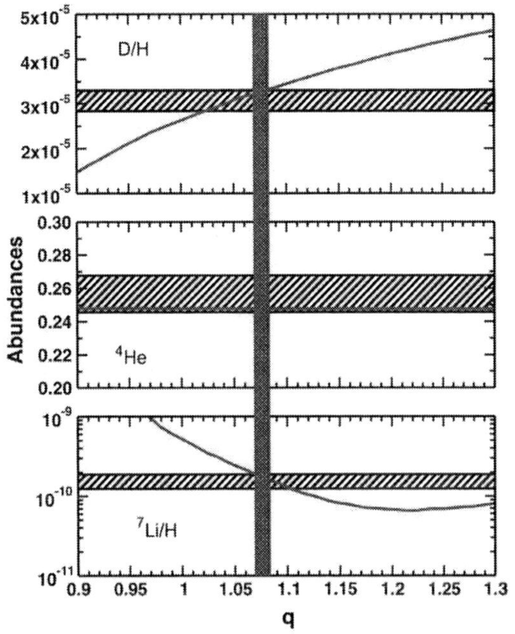

Figure 3. *Left:* Predicted abundances D, ^{4}He and ^{7}Li (red curves) as a function of the Tsallis parameter q [53]. The observed primordial abundances including 1σ uncertainties are indicated by hatched horizontal bands [8, 54, 55]. The vertical (blue) band refers to the parameter q within the interval $1.069 < q < 1.082$.

[45–47]. In Ref. [48], one of these non-extensive statistics, namely, the Tsallis statistics [46, 47] has been used to describe the relative velocities of particles during the BBN. The effect on the lithium abundance was again the motivation for this work. The Tsallis statistics was used [46, 47], because it represents a family of entropies depending on a parameter q, which measures the departure from Boltzmann statistics. The Boltzmann statistics is recovered when $q = 1$.

In all previous applications of non-extensive statistics, it has been found that the non-extensive parameter q does not depart appreciably from the Boltzmann value $q = 1$. Non-extensive Maxwellian velocity distributions have previously been applied to study stellar nuclear burning, e.g. in Refs. [49–52]. In Ref. [48] the Tsallis statistics was used to deduce reaction rates during the BBN and predictions were made for the ^{4}He, D, ^{3}He, and ^{7}Li abundances which are based on the reaction rates for p(n,γ)d, d(p,γ)^{3}He, d(d,n)^{3}He, d(d,p)t, ^{3}He(n,p)t, t(d,n)^{4}He, ^{3}He(d,p)^{4}He, ^{3}He(α,γ)^{7}Be, t(α,γ)^{7}Li, ^{7}Be(n,p)^{7}Li and ^{7}Li(p,α)^{4}He and their available experimental data [48]. The conclusion from Ref. [48] is that if either $q > 1$ of $q < 1$, the abundances of all elements are affected but that of ^{7}Li always increases. Therefore, it was inferred that the lithium problem always seems to get worse with the use of the Tsallis statistics.

However, there was a small, but relevant point neglected in the calculations of Ref. [48], namely, the proper inclusion of the reaction Q-values in the reaction rates obtained with the Tsallis statistics. This was fixed in Ref. [53] and shown that, when the Q-values for the reverse reactions are properly accounted for, a beautiful result emerges for a relatively small departure of the parameter q from unity. The abundances of H, D, ^{3}H, ^{3}He, and ^{4}He, do not change, but that of ^{7}Li does change appreciably, and in the correct direction to solve the ^{7}Li puzzle. In fact, an excellent agreement was found between the calculated and the primordial abundances observed for D, ^{4}He, and ^{7}Li for $1.069 < q < 1.082$, indicating that a possible solution to the cosmological lithium problem might arise from a fine tuning of the physics involved. This is shown in Figure 3 with the predicted abundances D, ^{4}He and ^{7}Li (red curves) as a function of the Tsallis parameter q. The observed primordial abundances including 1σ uncertainties are indicated by hatched horizontal bands [8, 54, 55]. The vertical (blue) band refers to the parameter q within the interval $1.069 < q < 1.082$.

The work published in Ref. [53] was cited as a research highlight by the American Astronomical Society [56]. It attests the relevance of the lithium puzzle and the anxiety that its solution entails for the astronomical community. The puzzle has been around the literature for a few decades already. The exercise played in Ref. [53] shows that a solution might be the outcome of a fine tuning of the physics during the BBN. The Tsallis statistics might be one possible departure from the standard physics during the big bang epoch. The question remains on the physical meaning for the value of $q \neq 1$, and its relation to other physical processes.

Acknowledgements
This work was supported in part by the U.S. DOE grant DE- FG02-08ER41533 and the U.S. National Science Foundation Grant No. 1415656.

References

[1] C.A. Bertulani and T. Kajino, Prog. Part. Nucl. Phys. 89, 56 (2016).
[2] E. Komatsu, et al., Ap. J. 192, 18 (2011).
[3] P.A.R. Ade, et al., Astron. Astrophys. 594, A13 (2016).
[4] The ALEPH Collaboration, The DELPHI Collaboration, The L3 Collaboration, The OPAL Collaboration, The SLD Collaboration, The LEP Electroweak Working Group, The SLD Electroweak and Heavy Flavor Groups, Phys. Reports 427, 257 (2006).
[5] C. Patrignani, et al. (Particle Data Group), Chinese Phys. C 40, 100001 (2016).
[6] M. Spite and F. Spite, Nature 297, 483 (1982); F. Spite and M. Spite, A&A 115, 357 (1982).
[7] W. Aoki, et al., Ap. J. 698, 1803 (2009).
[8] L. Sbordone, et al., A&A 522, A26 (2010).
[9] A. Coc, et al., Ap. J. 744, 158 (2012).
[10] R.G. Pizzone, et al., Ap. J. 786, 112 (2014).
[11] C.A. Bertulani, Comput. Phys. Commun. 156, 123 (2003).
[12] A. M. Mukhamedzhanov, Shubhchintak, and C.A. Bertulani, Phys. Rev. C 93,045805 (2016).
[13] M. Anders et al., Phys. Rev. Lett. 113, 042501 (2014).
[14] M. Asplund, D. L. Lambert, P. E. Nissen, F. Primas and V. V. Smith, Ap. J. 644, 229 (2006).
[15] K. Lind, J. Melendez, M. Asplund, R. Collet and Z. Magic, A&A 554, A96 (2013).
[16] R.V. Wagoner, Ap. J. Suppl. Ser. 18, 247 (1969).
[17] L. Kawano, "Let's Go: Early Universe. Primordial Nucleosynthesis: The Computer Way"; NASA Technical Reports Server (NTRS): Hampton, VA, USA, 1992.
[18] Y.I. Izotov and T.X. Thuan, Ap. J. Lett. 710, L67 (2010).
[19] E.E. Salpeter, Aust. J. Phys. 7, 373 (1954); E.E. Salpeter and H.M. Van Horn, Ap. J. 155, 183 (1969).
[20] B. Wang, C.A. Bertulani, and A.B. Balantekin, Phys. Rev. C 83, 018801 (2011).
[21] C. Spitaleri, et. al., Phys. Lett. B 755, 275 (2016).
[22] Signe Riemer-Sorensen and Espen Sem Jenssen, Universe 2, 44 (2017). doi:10.3390/universe3020044.

[23] R.J. Cooke, M. Pettini, and C.C. Steidel, arXiv:1710.02730.
[24] T. M. Bania, R. T. Rood, D. S. Balser, Nature 415, 54 (2002).
[25] C. Broggini, et al., JCAP 6, 30 (2012).
[26] S.Q. Hou, et al. 2015, Phys. Rev. C 91, 055802 (2015).
[27] M. Barbagallo, et al., Phys. Rev. Lett. 117, 152701 (2016).
[28] T. Kawabata, et al., Phys. Rev. Lett. 118, 052701 (2017).
[29] L. Lamia, et. al., Ap. J. 850, 175 (2017).
[30] M. Hartos, et. al., to be published.
[31] J. L. Feng, arXiv:1003.0904 (2010).
[32] G. Bertone, D. Hopper, and J. Silk, Phys. Rep. 405, 279 (2005).
[33] G. Bertone, Nature 468, 389 (2010).
[34] T.D. Li and C. N Yang, Phys. Rev. 104, 254 (1956).
[35] Y. Kobzarev, L. Okun, I. Pomeranchuk, Yad. Fiz. 3, 1154 (1966).
[36] M. Pavsic, Int. J. Theor. Phys. 9, 229 (1974).
[37] R. Foot, H. Lew, R. Volkas, Phys. Lett. B 272, 67 (1991).
[38] E. Akhmedov, Z. Berezhiani, G. Senjanovic, Phys. Rev. Lett. 69, 3013 (1992).
[39] O. Oliveira, et al., arXiv:1108.2723 (2011).
[40] C.A. Bertulani, et. al., AIP Proc.1498, 134 (2013).
[41] O. Oliveira, et al., Braz. J. Phys. 46, 721 (2016).
[42] Z.G. Berezhiani, A.D. Dolgov, R.N. Mohapatra, Phys. Lett. B375, 26 (1996).
[43] K.A. Olive (Particle Data Group - Review of Particle Physics), Chin. Phys. C 38, 090001 (2014).
[44] C.A. Bertulani, V. Challa, J.J. He, S.Q. Hou, Ravinder Kumar, arXiv:1704.07803.
[45] A. Rényi, Proc. of the 4th Berkeley Symposium on Mathematics, Statistics and Probability, p. 547, 561 (1960).
[46] C. Tsallis, J. Stat. Phys. 52, 479 (1988).
[47] M. Gell-Mann and C. Tsallis, eds., "Nonextensive Entropy-Interdisciplinary Applications, (Oxford University Press, New York, 2004). For a regularly updated bibliography see URL: http://tsallis.cat.cbpf.br/biblio.htm
[48] C.A. Bertulani, J. Fuqua and M.S. Hussein, Ap. J. 767, 67 (2013).
[49] M. Lissia and P. Quarati, Europhys. News 36, 211 (2005).
[50] J. Haubold and D. Kumar, Astroparticle Phys. 29 ,70 (2008).
[51] S. Degl'Innocenti et al, Phys. Lett. B 441, 291 (1998).
[52] M. Coraddu, et al., Braz. J. Phys. 29, 153 (1999).
[53] S.Q. Hou, et. al., Ap. J. 834, 165 (2017).
[54] E. Aver, K.A. Olive, E.D. Skillman, JCAP 5, 003 (2010).
[55] K.A. Olive, P. Petitjean, E. Vangioni, J. Silk, MNRAS 426, 1427 (2012).
[56] Susanna Kohler, AAS-NOVA, "Fixing the Big Bang Theory's Lithium Problem", 15 February (2017).

The LUNA experiment: past and future

Carlo Broggini[1,⋆]

[1] *INFN-Padova, via F.Marzolo 8, I-35131 Padova*

Abstract. The essential ingredients of nuclear astrophysics are the thermonuclear reactions which shape the life and death of stars and which are responsible for the synthesis of the chemical elements in the Universe. Deep underground in the Gran Sasso Laboratory the cross sections of the key reactions responsible for the hydrogen burning in stars have been measured with two accelerators of 50 and 400 kV voltage right down to the energies of astrophysical interest. In particular, the main results obtained during the 'solar' phase of LUNA are here reviewed and their influence on our understanding of the properties of the neutrino and of the Sun is discussed. Then, the future of LUNA during the next decade is outlined. It will be mainly focused on the study of the nuclear burning stages after hydrogen burning: helium and carbon burning. All this will be accomplished thanks to a new 3.5 MV accelerator able to deliver high current beams of proton, helium and carbon which will start running under Gran Sasso in 2019.

1 Introduction

Stars are not perfect and eternal bodies as believed by the ancient philosophers. On the contrary, gravity triggers the birth of a star which then evolves as a fusion reactor to finally die out in a quiet or violent way, depending on its initial mass. Only hydrogen, helium and lithium are synthesized in the first minutes after the big-bang. Almost all the other elements in the Universe are produced in the reactions taking place inside the 'cauldrons' active in the cosmos, i.e. the stars. Nuclear astrophysics studies the reactions which transmute the chemical elements and, by doing so, supply stars with the power to shine. The key point of a nuclear reaction is the value of its cross section at the energy at which the reaction takes place.

In particular, the reaction occurs in the hot plasma of a star, with temperatures in the range of tens to hundreds of millions degrees, corresponding to an energy window, the Gamow peak, which is far below the Coulomb energy arising from the repulsion between nuclei. At the Gamow peak the cross section can be approximated as:

$$\sigma(E) = \frac{S(E)}{E} exp(-2\pi\eta), \tag{1}$$

where $S(E)$ is the astrophysical factor (which contains the nuclear physics information) and η is given by $2\pi\eta = 31.29\, Z_1\, Z_2(\mu/E)^{1/2}$. Z_1 and Z_2 are the nuclear charges of the interacting particles, μ is the reduced mass (in units of amu), and E is the center of mass energy (in units of keV).

The extremely low value of the cross section at the stellar energies, ranging from pico to femto-barn and even below, has always prevented its measurement in a laboratory at the Earth's surface,

⋆e-mail: broggini@pd.infn.it, for the LUNA Collaboration

where the signal to background ratio is too small mainly because of cosmic ray interactions. In order to explore the low energy domain of nuclear astrophysics LUNA (Laboratory for Underground Nuclear Astrophysics) started its activity in 1991 as a pilot project with a 50 kV electrostatic accelerator installed inside the laboratory under Gran Sasso (Fig.1), to reproduce in the laboratory what Nature makes inside the stars (for exhaustive reviews refer to [1–3]).

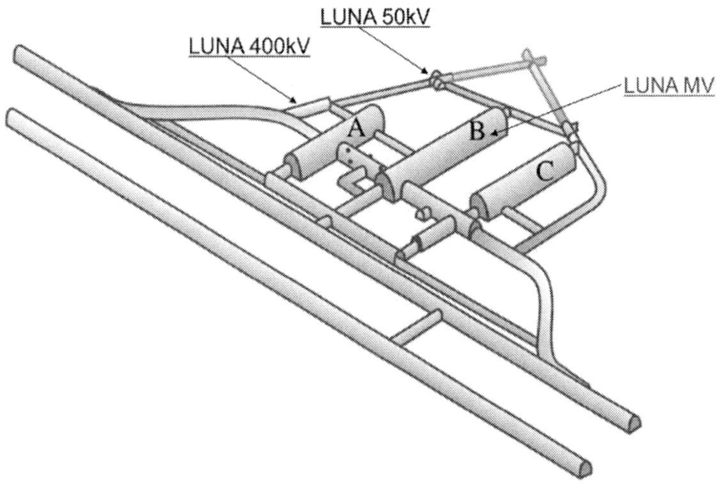

Figure 1. The underground Gran Sasso Laboratory, with the location of the three different LUNA accelerators.

2 Background and accelerators under Gran Sasso

The dolomite rock of Gran Sasso provides a natural shield equivalent to at least 3800 meters of water which reduces the muon and neutron fluxes by a factor 10^6 and 10^3, respectively. The remaining neutron flux is mainly due to (α,n) reactions in the rock, with the alpha particles coming from the ^{238}U and ^{232}Th decay chains. As a consequence, both the prompt and delayed activity from muon and neutron interactions are strongly suppressed. This way the γ-rays due the ^{238}U and ^{232}Th chains and ^{40}K in the rock can be removed by a suitable passive shield surrounding the target and the detectors (usually low-background lead and copper), without having the shield itself activated because of cosmic ray interactions. Finally, the activity due to Radon emanating from the rock is suppressed down to the level of few tens of Bequerel/m^3 thanks to frequent air volume exchanges (once every 3.5 hours). Eventually, the background spectrum of a 'clean' γ-ray detector can be reduced by 4-5 orders of magnitude underground as compared to the overground. Not so large is the background suppression exhibited by alpha particle detectors running underground, about a factor of 15, but still crucial for some measurements.

The accelerators used at LUNA are characterized by: high current, in the range of several hundreds μA, mandatory because of the small value of the cross section to be measured, acceleration potential stable to $\leq$1 kV over many hours in order to avoid changes in the yield due to the Coulomb barrier, and to $\leq$0.1 kV over one hour, in order to reliably perform energy scans of targets, with the beam energy spread on a similar level ($\leq$1 keV). Two ion accelerators have hitherto been used at LUNA, and a third one will be installed soon. All three are of the electrostatic type, and the high voltage is supplied by Cockroft-Walton generators. The first accelerator, the LUNA 50 kV machine, was operated with a duo-plasmatron ion source for intense and well focused hydrogen and helium beams. The main problem of this source is the relatively short lifetime. In order to solve this issue, for the LUNA 400 kV accelerator, in operation since 2001, a radio-frequency ion source is used to provide hydrogen or helium ions. The whole accelerator is enclosed in a pressure tank filled with 20 bar of a gas mixture consisting of nitrogen and carbon dioxide.

The third LUNA accelerator, LUNA MV, will be installed at the end of this year. Briefly, the acceleration potential of 3.5 MV is an order of magnitude larger than that of the LUNA 400 machine. The new accelerator will be able to provide not only intense $^{1}H^{+}$ and $^{4}He^{+}$ beams, but also carbon beams (single and double-charged, with particle current of the order of 100 and 50 μA, respectively). After the accelerator, the ion beam passes through an electromagnet used as a beam analyzer. The LUNA MV accelerator, as the LUNA 400 kV one, will be equipped with two different beam lines, only one of them fed at a time. The accelerator room, with a surface of about 300 m^2, will have 80 cm thick concrete walls and ceiling working as neutron shielding. The maximum neutron flux just outside the concrete walls will be significantly smaller than the natural neutron background at LNGS, with a similar energy spectrum.

3 The solar phase

The measurement of the $^{3}He(^{3}He,2p)^{4}He$ cross section within the solar Gamow peak (16-27 keV) has been the reason why LUNA was born in 1991. Such a reaction is a key one of the hydrogen burning proton-proton chain, which is responsible for more than 99% of the solar luminosity. A resonance in its cross section at the thermal energy of the Sun was suggested at the beginning of the seventies to explain the low measured solar neutrino flux, as compared to the Standard Solar Model predictions.

The final experimental set-up for the study of $^{3}He(^{3}He,2p)^{4}He$ was made of eight 1 mm thick silicon detectors of 5x5 cm^2 area placed around the beam inside the windowless target chamber filled with ^{3}He at the pressure of 0.5 mbar. The simultaneous detection of two protons has been the signature which unambiguously identified the $^{3}He(^{3}He,2p)^{4}He$ fusion reaction. For the first time a nuclear reaction has been measured in the laboratory at the energy occurring in a star [4]. In particular, at the lowest energy of 16.5 keV the cross section is 0.02 pb, which corresponds to a rate of about 2 events/month, rather low even for the "silent" experiments of underground physics. No narrow resonance has been found and, as a consequence, the astrophysical solution of the ^{8}B and ^{7}Be solar neutrino problem based on its existence was definitely ruled out.

$^{3}He(\alpha,\gamma)^{7}Be$, the competing reaction for ^{3}He burning, has also been measured by LUNA both by detecting the prompt γ rays and by counting the ^{7}Be nuclei from their decay. The two different methods gave the same result within the total error of 4% [5].

The third cornerstone provided by LUNA to the Standard Solar Model has been the cross section of $^{14}N(p,\gamma)^{15}O$ at low energy. $^{14}N(p,\gamma)^{15}O$ is the slowest reaction of the first CNO cycle, it rules its energy production rate and it determines the ^{13}N and ^{15}O solar neutrino flux, which depends almost linearly on the $^{14}N(p,\gamma)^{15}O$ cross section. The results obtained by LUNA first with a germanium detector [6] and then with a high efficiency BGO set-up [7] were about a factor of two lower than the existing extrapolation from previous data. On the other hand, they were in good agreement with the results obtained with indirect methods [8]. Because of this reduction the predicted CNO neutrino yield in the Sun was decreased by about a factor of two and the lower limit on the age of the Universe inferred from the age of the oldest stellar populations, the globular clusters, was increased by 0.7-1 billion years [9] up to 14 billion years.

We point out that the total error on the extrapolated S-factor has been reduced by LUNA to 8% (when we started it was 50%): $S_{1,14}(0)$=1.57±0.13 keVb [10]. Such an error is small enough to allow for the extraction of the carbon and nitrogen content of the Sun core from the forthcoming measurement of the CNO solar neutrinos [11]. This way it will be possible to test whether the early Sun was chemically homogeneous, a key assumption of the Standard Solar Model.

4 The first 5 years of LUNA MV

With the new 3.5 MV accelerator, planned to start running underground in Hall B next year (Fig.2), LUNA will enter a new phase, going to last more than 10 years. Here we describe the scientific program related only to the first 5 years of activity as given in the LUNA MV proposal [12]. The first

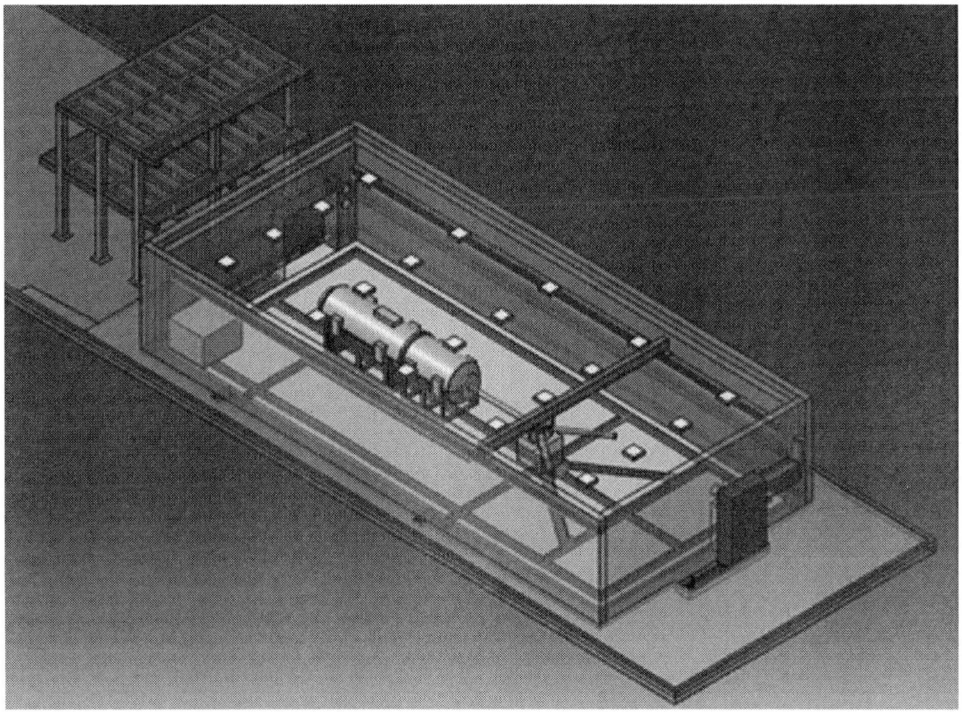

Figure 2. The LUNA MV experimental hall in the North side of Gran Sasso Hall B. The structure to the left will host counting room and ancillary facilities.

reaction to be studied will be $^{14}N(p,\gamma)^{15}O$, already measured with the 400 kV accelerator. This way we will perform the tuning of LUNA MV verifying the performances of the accelerator and of the experimental setup. In addition, a measurement over a wide energy region can provide valuable data to reduce the 8% error in the low energy extrapolation of the cross section. Such a measurement is particularly required now since a new result, obtained by using a 1 MV and a 4 MV accelerators at the University of Notre Dame, reported some inconsistencies for the ground state transition. Then, we will focus the activity of one of the two beam lines on the study of $^{12}C+^{12}C$: this measure will be the flagship of the first 5 years of LUNA MV. Alternating in time with $^{12}C+^{12}C$, the study of $^{13}C(\alpha,n)^{16}O$ and $^{22}Ne(\alpha,n)^{25}Mg$, the two reactions responsible for neutron generation in stars, will be performed on the other beam line.

4.1 ^{12}C+^{12}C

The end of helium burning marks a branching point in stellar evolution. Lower mass stars will become stable, electron-degenerate white dwarfs, while higher mass stars will enter the quiescent carbon burning phase of their evolution. This mass limit, $\sim 9M_{\odot}$, which separates the progenitors of white dwarfs, novae and type Ia supernovae, from those of core-collapse supernovae, neutron stars, and stellar mass black holes, is given by the ^{12}C+^{12}C reaction rate in the temperatures range between 0.5 and 1 GK (center of mass energies between 0.9 and 3.4 MeV). ^{12}C+^{12}C is also affecting nucleosynthesis in massive stars. The ^{12}C+^{12}C reaction, characterized by a Coulomb barrier of about 6.7 MeV, mainly proceeds through two different channels corresponding to the emission of a proton or an α particle. The Q-value for proton emission is 2.24 MeV while that for α emission is 4.62 MeV. The proton and alpha channels can be measured either by detecting the charged particles or by revealing the gamma decay of the first excited state to the ground state of the ^{23}Na (440 keV) or ^{20}Ne (1634 keV) residual nuclei, respectively. Approximately, the decay of the first excited state to the ground state accounts for 50% of the total cross section.

So far, many different experiments attempted to measure the ^{12}C+^{12}C cross section ([12] and references therein). The lowest energy reached in a direct experiment is 2.1 MeV, with hydrogen and deuterium contaminations inside the carbon target responsible for the most severe beam induced background. In LUNA MV a detection system made of a high efficiency and ultra low intrinsic background HPGe detector complemented with silicon detectors (or telescopes) at backward angles may allow the measurement of both the γ-rays and the particles (protons and alphas). The deep underground location of the accelerator and its capability of producing an intense carbon beam offer a unique opportunity to perform such a measurement entering for the first time the low energy region below 2 MeV.

5 The neutrons for the s-process

The neutrons for the astrophysical s-process are supplied by two (α,n) reactions: At low temperatures, $\sim$ 90 MK, ^{13}C(α,n)^{16}O plays the major role. It operates in the He-rich shell of low-mass (less than 4 solar masses) AGB stars, the so called intershell, between the helium and hydrogen burning shells. It is the neutron source reaction that allows the creation of the majority of the s-process elements. At higher temperatures, $\sim$ 300 MK, ^{22}Ne from one or two helium capture processes on ^{18}O or ^{14}N is the seed for the ^{22}Ne(α,n)^{25}Mg reaction. ^{22}Ne(α,n)^{25}Mg operates in the He-burning shell of high-mass (more than 4 solar masses) AGB stars and during the core-He burning and the shell-C burning of massive stars (more than 10 solar masses). The latter is responsible for the synthesis of the s-process elements with mass number A smaller than 90.

The accurate and precise (at the level of 10%) knowledge of the ^{13}C(α,n)^{16}O reaction is required at stellar temperatures in the range from 80 to 250 million K (below about 230 keV center of mass energy). The current experimental situation does not fulfil such a requirement (see [12] and references therein). In particular, there exist no data from direct measurements close to the energy of astrophysical interest, the lowest energy data are affected by uncertainties that are too large to constrain extrapolations and, finally, discrepancies exist between different data sets both in energy dependence and absolute values. LUNA will study ^{13}C(α,n)^{16}O, Q=2.216 MeV, in direct kinematics both with the 400 kV accelerator and then with the 3.5 MV one. This way it will be possible to cover a wide energy range, to address the issue of normalization discrepancies and to minimize overall statistical and systematic uncertainties, taking advantage of the 3 orders of magnitude suppression of the laboratory neutron background. Given the low background, it may be possible to approach center of mass energies in the 200 keV range. Finally, we point out that the same energy range of interest can be covered in inverse kinematics using a ^{13}C beam with energies $E = 0.9 - 4.5$ MeV. A ^{4}He gas target would be one possible solution, significantly changing the systematic uncertainties and background sources as compared to the direct kinematics measurement.

The current knowledge of the $^{22}Ne(\alpha,n)^{25}Mg$ reaction is also incomplete and imprecise ([12] and references therein). Current estimates of the rate are mostly based on experimental evaluations of the dominant resonance at 832 keV and provide the rate with an uncertainty of 20-30%, while less than 5% is required for accurate model predictions. Furthermore, extrapolations to low energies may be affected by the unknown influence of low-energy resonances just below the neutron threshold, casting doubts on the accuracy of the values currently adopted in the stellar models. In LUNA MV it will be possible to study $^{22}Ne(\alpha,n)^{25}Mg$ (Q=-478 keV) with the α beam on a windowless gas target of enriched ^{22}Ne surrounded by a neutron detector. It may be feasible to have an accurate cross section measurement down to the center of mass energy in the 600 keV range. At lower energies it will be possible to exclude any resonance able to give a significant contribution to neutron production in AGB stars.

6 Conclusions

In 1991 LUNA started underground nuclear astrophysics in the core of Gran Sasso, below 1400 meters of dolomite rock. The extremely low background has allowed for nuclear physics experiments with very small count rate, down to a few events per year. The important reactions responsible for the hydrogen burning in the Sun have been studied for the first time down to the relevant energies, providing fundamental contributions for the prediction of the solar neutrino spectrum. Next year LUNA will enter a new phase: the study of helium and carbon burning with a 3.5 MV accelerator able to deliver high current beams of hydrogen, helium and carbon. The program of the first five years of running time is now supposed to be focused on the study of $^{12}C+^{12}C$, the reaction which is switching on carbon burning. The other two key reactions which will be studied during this phase are $^{13}C(\alpha,n)^{16}O$ and $^{22}Ne(\alpha,n)^{25}Mg$, responsible for generating free neutrons inside stars. These neutrons give then rise to the s-process responsible for the synthesis of about half of the chemical elements beyond the iron peak in the Universe.

References

[1] H.Costantini, A.Formicola, G.Imbriani, M.Junker, C.Rolfs and F.Strieder, Rep. Progr. Phys. **72**, 086301 (2009)
[2] C.Broggini, D.Bemmerer, A.Guglielmetti and R.Menegazzo, Ann. Rev. Nucl. Part. Science **60**, 53 (2010)
[3] C.Broggini, D.Bemmerer, A.Caciolli and D.Trezzi, Progr. Part. Nucl. Phys. **98**, 55 (2018)
[4] R.Bonetti et al., Phys. Rev. Lett. **82**, 5205 (1999)
[5] F.Confortola et al., Phys. Rev. C **75**, 065803 (2007)
[6] A.Formicola et al., Phys. Lett. B **591**, 61 (2004)
[7] A.Lemut et al., Phys. Lett. B **634**, 483 (2006)
[8] A.Mukhamedzhanov et al., Phys. Rev. C **67**, 065804 (2003)
[9] G. Imbriani et al., Astron. Astrophys. **420**, 625 (2004)
[10] M.Marta et al., Phys. Rev. C **78**, 022802 (2008)
[11] A.Serenelli, Eur. Phys. J. A **52**, 78 (2016)
[12] LUNA MV Proposal, **INFN-17-19/LNGS** (2017) in http://www.lnf.infn.it/sis/preprint/search.php

The s-process in stellar sites

Sergio Cristallo[1,2,★]

[1] *Osservatorio Astronomico d'Abruzzo, via Maggini, 64100 Teramo, Italy*
[2] *Sezione INFN di Perugia, via Pascoli, 06123 Perugia, Italy*

Abstract. Stars are marvellous caldrons where all the elements of the Universe (apart from hydrogen and helium) have been synthesized. The solar system chemical distribution is the result of many pollution episodes from already extinct stellar generations, occurred at different epochs before the Sun formation. Main nucleosynthesis channels responsible for the formation of heavy elements are the rapid neutron capture process (the r-process) and the slow neutron capture process (the s-process). Hereafter, I will describe the theory of the s-process and the stellar sites where it is active.

1 Introduction

The history of the chemical enrichment of the Universe follows a simple paradigm: helium and metals (i.e. elements with A≥12) increase at the expense of hydrogen. Starting with an hydrogen mass fraction of about 0.25 and no metals, the Universe evolved to the current ~65% of H and ~4% of metals characterizing star forming regions. The distribution of heavy elements (A>56) in the solar system shows an exponential decline, together with clearly identified abundance peaks, related to the s (slow) and to the r (rapid) processes [1]. The abundances of heavy elements are far greater that can be reproduced in nuclear statistical equilibrium, so that they seem to require a non-equilibrium mechanism. Their synthesis cannot occur via charged particle reactions, due to the increasing Coulomb barrier with increasing atomic number. Moreover, ^{56}Fe is the most tightly bound nucleus and, therefore, neither fusion nor fission processes may release energy starting from it. Thus, the only way to efficiently produce isotopes heavier than iron is via neutron capture processes. In such a case, in fact, heavy elements can be easily created, even at extremely low energies (due to the fact that, in general, neutron capture cross sections increase with decreasing energy).
Multiple neutron captures on a nucleus (Z,A) occur if the new synthesized neutron-rich isotopes (Z,A+i) are stable (with i ranging from 1 to 6). On the contrary, if the nucleus (Z,A+i) is radioactive, the question whether it will decay to its isobar (Z+1,A+i) or will capture another neutron depends upon the relative lifetimes between the decay lifetime and the timescale of the neutron capture process. One might expect to find in stars a large spread of neutron fluxes, each one characterizing different evolutionary phases. On the contrary, the most common fluxes are either quite small (s-process: $n_n \sim 10^7$ cm^{-3}) or quite large (r-process: $n_n > 10^{20}$ cm^{-3}). The r process is believed to occur during late phases of massive stars evolution, even if the exact site is still matter of debate (SuperNovae [2, 3] vs. Neutron stars mergers [4]). Due to the difficulties in modelling such a process, together

★e-mail: sergio.cristallo@inaf.it

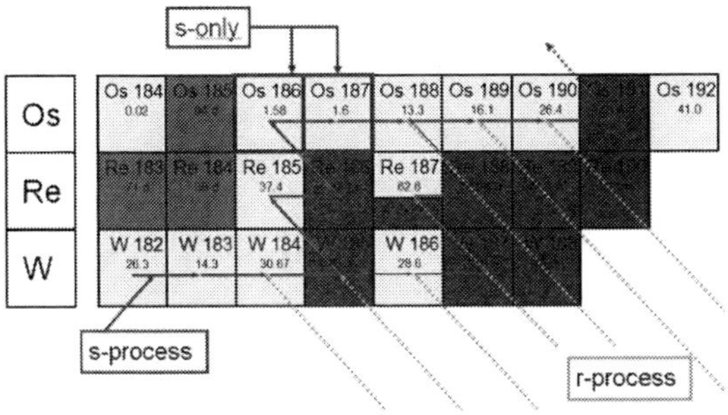

Figure 1. Scheme of the s-process pattern in the W-Re-Os region.

with the still large uncertainties affecting its nuclear inputs, the contribution to the Sun provided by the r-process is commonly estimated by subtracting that coming from the s-process (r=1-s). A correct estimate of s-process yields is thus mandatory for a detailed knowledge of the chemical evolution of our Galaxy.

2 The theory of the s-process

A typical s-process pattern is reported in Figure 1. The Re-Os-W region is shown. Stable isotopes are colored in gray, while unstable isotopes are marked with different colors (red for those decaying β^+ or capturing an electron; blue for those decaying β^-). Numbers in each box represent the solar isotopic fraction (for stable isotopes) and decay lifetimes (for the unstable ones). The main s-process flow is traced by red heavy arrows (horizontal lines are neutron captures, while slanting ones are β^- decays). In correspondence of unstable isotopes (e.g. ^{185}W), s-process branchings may appear and two different paths may develop. The predominance of the channel is determined by the ratio between the neutron capture cross section timescale and the decay lifetime. The contribution from the r-process is highlighted by dotted green arrows. Interestingly, there are isotope which are shielded by an r-process contribution by their stable isobars (see, for instance, the couples ^{186}Os-^{186}W and ^{187}Os-^{187}Re). Those isotopes are named s-only. Their nucleosynthesis is of paramount importance to calibrate s-process calculations (see [17] and references therein). Main seeds for the s-process are ^{56}Fe nuclei. In Figure 2 the solar abundances for isotopes heavier than iron are shown (upper panel). Three peaks clearly emerge from the distribution, corresponding to nuclei with neutron magic nuclei (N=50, N=82 and N=126). Those isotopes are characterized by closed shells configurations and, thus, they are particularly stable with respect to their neighbors. This can be better understood looking at the lower panel of Figure 2, where neutron capture cross sections are reported (in mb). Among the isotopes with the slowest neutron capture cross sections there are ^{88}Sr, ^{138}Ba and ^{208}Pb. They characterize the three peaks of the s-process: *ls* elements (Sr-Y-Zr), *hs* elements (Ba-La-Ce-Pr-Nd) and lead (Pb). The precise knowledge of their neutron capture cross sections is of paramount importance to determine the ratio between light and heavy s-process elements, the so called [*hs*/*ls*] s-process index. For this reason, dedicated experimental campaigns are currently going on at the n_TOF facility at CERN [7]. At atomic weights far from the closed shells, the so-called *local approximation* holds, i.e. the product σ·n is almost constant. In this case, the cross sections are so large that the differences between the products is much smaller than the magnitude of either one of them.

Another important quantity in s-process calculations is the neutron-to-seed ratio, i.e. the ratio between the available neutrons and the seeds from which the s-process start. Depending on this ratio, different s-process distributions can be obtained. In fact, the larger the number of available neutrons per iron seed, the larger the masses of the synthesized isotopes. For instance, at solar metallicity only *ls* elements are efficiently produced, due to the very low neutron cross sections of ^{88}Sr, ^{89}Y and ^{90}Zr,

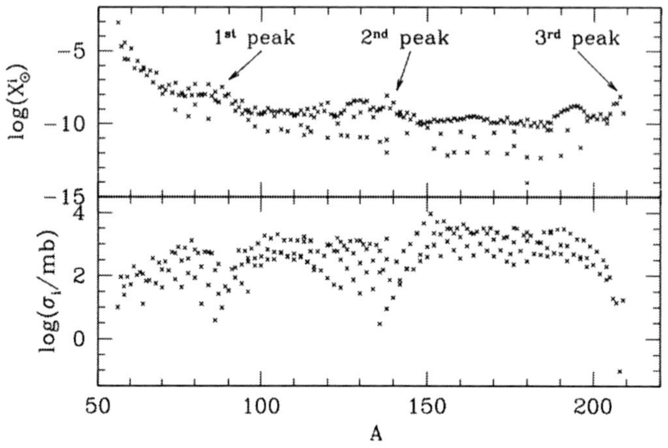

Figure 2. Heavy elements isotopic solar distribution (upper panel) and corresponding neutron capture cross sections (lower panel). Experimental data are mainly taken from [6].

which act as bottlenecks for the s-process path. If such a barrier is overtaken, the next one is encountered at larger atomic masses (i.e. for ^{139}Ba, ^{139}La and ^{140}Ce). At low metallicities the s-process pattern goes beyond this second barrier and lead can be largely synthesized. The different s-process trends as a function of the initial metallicity are reported in the left panel of Figure 3.
Major neutron sources in stars are the $^{13}C(\alpha,n)^{16}O$ and the $^{22}Ne(\alpha,n)^{25}Mg$ reactions. While the number of iron seeds directly scales with metallicity, the number of neutrons may have a primary behavior, that is it may be a direct product of H- and He-burning. In such a way, different s-process distributions result for different initial metallicities. The first reaction dominates in low mass stars (M< 4 $M_\odot$, see Section 3) and it is responsible for the *main* component of the s-process (88≤A≤208). The second, instead, is the main neutron producer in massive stars (see Section 4) and it accounts for the *weak* component of the s-process (A<88). Due to their importance in determining the efficiency of the s-process nucleosynthesis, both reactions have been explored into detail in the past (see [10] and [11], respectively).

3 The s-process in low and intermediate mass stars

The s process is at work in Thermally Pulsing AGB (TP-AGB) phase of low mass stars ($1 < M/M_\odot < 4$; [12, 13]). In those objects, freshly synthesized elements are carried out to the surface by means of recurrent mixing episodes, commonly known as Third Dredge Ups (TDUs) (see [9] and references therein; see also [14]). Interested readers may download TP-AGB surface abundances and yields from the FRUITY database[1] [15, 16]. The structure of an AGB star is made of a partially degenerate carbon-oxygen core, an He shell separated from an H shell by a thin region, and by a convective envelope. The energy is provided for most of the time by the H-shell, located below the H-rich envelope. Recurrently, the H-burning in the shell is interrupted by the growing up of thermonuclear runaways, triggered by violent He-burning ignitions (Thermal Pulses, TPs). During a TP, the region between the two shells becomes convective, while the layers above expand. As a consequence, the

[1] http://fruity.oa-teramo.inaf.it/

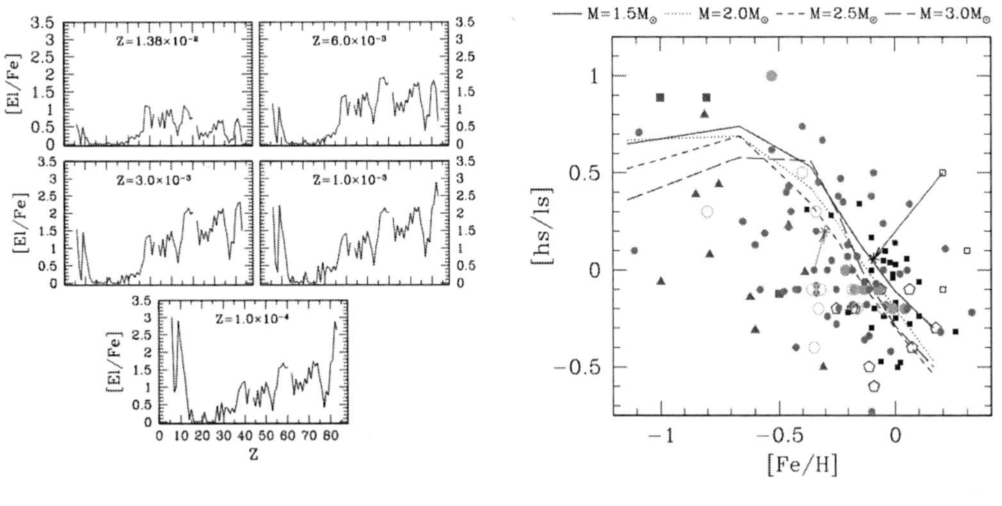

Figure 3. Left panel: s-process distributions at different metallicities for models with initial mass $M=2M_\odot$ (from [8]). Right panel: comparison between theory and observations at different metallicities (from [15]).

H-burning shell possibly switches off, allowing the convective envelope to penetrate in regions where heavy isotopes have been freshly synthesized. As already highlighted, the main neutron sources is the $^{13}C(\alpha,n)^{16}O$ reaction. In order to obtain a sufficient amount of ^{13}C for the activation of the s-process, protons from the envelope have to be mixed (with a low efficiency) with the ^{12}C-rich material in the underlying radiative zone. Those protons are captured by ^{12}C and form ^{13}C via the nuclear chain $^{12}C(p,\gamma)^{13}N(\beta^-)^{13}C$. In such a way, a ^{13}C-enriched layer, the so-called ^{13}C-pocket, forms. The ^{13}C burns in radiative conditions between two TPs at temperatures around 100 MK. The attained neutron densities are $\sim 10^7$ cm^{-3}. The physical process leading to the formation of the ^{13}C pocket is far from being understood. In the past, different mechanisms have been proposed, among which opacity-induced mixing [8], Kelvin-Helmholtz instabilities plus gravity waves [18] and magnetic buoyancy [20]. Other non-convective mixings, such as those induced by rotation (e.g. Eddington-Sweet circulations and Goldreich-Schubert-Fricke instabilities), have been considered by [19], who demonstrated their relevance in shaping the ^{13}C distribution in the pocket (with obvious consequences on the following s-process nucleosynthesis). In the left panel of Fig. 3 typical s-process distributions as a function of the initial metallicity are reported. The showed patterns, which have been obtained with the FUNS code [8, 9], highlight the strong dependence of s-process distributions from the initial iron content. In fact, as soon as the latter decreases, the abundances of *hs* elements increase, followed by a rapid grow of the lead peak at very low metallicities. Note also the progressive growth with decreasing metallicities of the overabundances of carbon and fluorine. As a further demonstration of the key role of the neutron-to-seed ratio, we report in the right panel of Fig. 3 a comparison between theory and observations, where we plot the [*hs*/*ls*] s-process index as a function of the initial iron content.

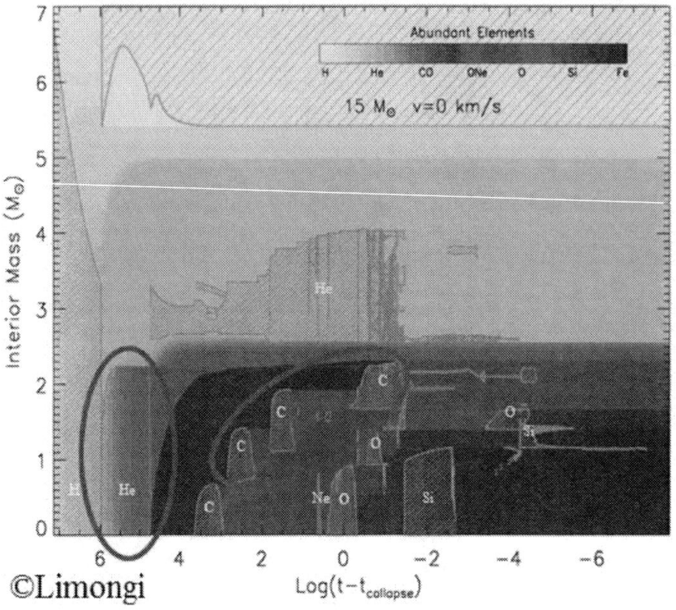

Figure 4. Kippenhahn diagram of a non rotating star with initial mass M=15 $M_\odot$ (copyright Marco Limongi).

Starting from the largest metallicities, it can be seen that the *[hs/ls]* ratio increases with decreasing metallicities, proving that heavier elements (in this case the *hs* ones) are produced. At [Fe/H]=-0.7 a plateau in the *[hs/ls]* appears. This metallicity marks the region where lead starts being efficiently synthesized, due to the strong decrease of the neutron-to-seed ratio.
In low mass AGB stars, an additional (marginal) contribution comes from the ^{22}Ne(α,n)^{25}Mg reaction, which burns at higher temperatures (T~ 300MK) during TPs. During this episode, neutron densities are larger (> 10^{11} cm^{-3}), but the neutron exposure (that is the integrated exposure to neutron fluence) is definitely lower with respect to the radiative ^{13}C burning. This derives from the fact that the neutron irradiation lasts for a lower time interval (yrs compared to tens of thousands yrs). The ^{22}Ne forms during TPs via two α captures: ^{14}N(α,γ)^{18}F(β^+)^{18}O(α,γ)^{22}Ne. This source is dominant in more massive AGBs (5< M/$M_\odot$ <8), because temperatures largely exceed 300 MK during TPs. Moreover, those objects have larger core masses and, therefore, the contribution from the ^{13}C is marginal [17]. Intermediate Mass AGB stars marginally contribute to the Galactic chemical evolution. Their role, instead, is of primary importance in the pollution history of s-rich galactic Globular Clusters (GCs), such as M22 [21] and M4 [22]. In fact, [23] (and independently [24]) demonstrated that AGB stars with masses between 3 and 6 $M_\odot$ have been the dominant polluters of those systems.

4 The s-process in massive stars

Elements between iron and the first s-process peak are efficiently synthesized by the weak s-process [25, 26]. This nucleosynthesis process is at work during the core He-burning and the C-shell burning phases of massive (M> 9 $M_\odot$) stars evolution (see Figure 4). In both episodes, neutrons are provided by the abundant ^{22}Ne, whose formation path is the same as in lower mass stars. During core He-burning, central temperatures exceeds 300 MK and efficiently activate the ^{22}Ne(α,n)^{25}Mg reaction. This phase, lasting about 1 Myr and characterized by rather low neutron densities (n~ 10^6 cm^{-3}),

almost yields the final weak s-process abundances of the model. During core He-burning, ^{22}Ne is not completely burnt. Thus, additional neutrons are available during the following C-burning shell phase[2]. In this phase, the α particles are provided by ^{12}C(^{12}C,α)^{20}Ne reactions. The neutron exposure is roughly the same as during core He-burning. However, this evolutionary phase is definitely shorter ($\sim$ 1 kyr) and the temperature definitely higher (T> 1GK). This leads to very large neutron densities (exceeding > 10^{12} cm^{-3}), but to a rather low neutron exposure. We refer to [27] for a detailed description of the nucleosynthesis and of the effects induced by variations of nuclear reaction rates.
Results are definitely different if the star is rotating. In that case, in fact, rotation induced mixing carries ^{12}C from the He convective core in the H burning shell, where it is transformed to ^{14}N via CNO burning. This has deep consequences on ^{22}Ne, whose abundance largely increases. The larger the initial rotation rate, the larger the ^{22}Ne production, with important consequences on the s-process production by massive stars at low metallicities [28].

References

[1] Burbidge, E. M. ,Burbidge, G. R., Fowler, W. A. & Hoyle, F., Rev. Mod. Phys. **29**, 547 (1955)
[2] Farouqi, K., et al., Astroph. Journ., **712**, 1359 (2010)
[3] Nishimura, N., et al., Astroph. Journ. Lett., **836**, 21 (2017)
[4] Kasen, D., et al., Nature, **551**, 80 (2017)
[5] Cristallo, S., et al., Astroph. Journ., **801**, 53 (2015)
[6] Bao, Z.Y. & Kaeppeler, F., Atom. Data Nucl. Data Tables, **76**, 70 (2000)
[7] Guerrero, C., et al., Eur. Phys. J. A, **49**, 27 (2013)
[8] Cristallo, S., et al., Astroph. Journ., **696**, 797 (2009)
[9] Straniero, O., et al., Nucl. Phys. A, **777**, 311 (2006)
[10] Trippella, O., et al., Astroph. Journ., **837**, 41 (2017)
[11] Massimi, C., et al., Phys. Lett. B, **768**, 1 (2017)
[12] Gallino, R., et al., Astroph. Journ., **497**, 388 (1998)
[13] Busso, M., et al., Ann. Rev. Astron. Astrophys., **37**, 239 (1999)
[14] Cristallo, S., et al., Nuovo Cimento B, **122**, 6 (2007)
[15] Cristallo, S., et al., Astroph. Journ. Suppl., **197**, 2 (2011)
[16] Cristallo, S., et al., Astroph. Journ., **833**, 181 (2016)
[17] Cristallo, S., et al., Astroph. Journ. Suppl., **219**, 21 (2015)
[18] Battino, U., et al., Astroph. Journ., **827**, 30 (2016)
[19] Piersanti, L., et al., Astroph. Journ., **774**, 98 (2013)
[20] Trippella, O., et al., Astroph. Journ., **818**, 125 (2016)
[21] Roederer, I., et al., Astroph. Journ., **742**, 27 (2011)
[22] Yong, D., et al., Astroph. Journ., **689**, 1031 (2008)
[23] Straniero, O., et al., Astroph. Journ., **785**, 77 (2014)
[24] Shingles, L.., et al., Astroph. Journ., **795**, 34 (2014)
[25] Prantzos, N.., et al., Astron. & Astrophys., **234**, 211 (1990)
[26] Chieffi, A. & Limongi, M., Astroph. Journ., **764**, 21 (2013)
[27] Pignatari, M., et al., Astroph. Journ., **710**, 1557 (2010)
[28] Prantzos, N., et al., MNRAS, **476**, 3432 (2018)

[2]Note that any nucleosynthesis occurring during the core C-burning phase will not produce chemical variations in the ejecta, because such a material will collapse in the central remnant (being a neutron star or a black hole).

The α-decay of the Hoyle state in ^{12}C: a new high-precision investigation

D. Dell'Aquila[1,2,⋆], *I.* Lombardo[4], *G.* Verde[4], *M.* Vigilante[1,2], *L.* Acosta[5], *C.* Agodi[6], *F.* Cappuzzello[7,6], *D.* Carbone[6], *M.* Cavallaro[6], *S.* Cherubini[6,7], *A.* Cvetinovic[6], *G.* D'Agata[7,6], *L.* Francalanza[2], *G.L.* Guardo[6], *M.* Gulino[8,6], *I.* Indelicato[6], *M.* La Cognata[6], *L.* Lamia[7], *A.* Ordine[2], *R.G.* Pizzone[6], *S.M.R.* Puglia[6], *G.G.* Rapisarda[6], *S.* Romano[6], *G.* Santagati[6], *R.* Spartà[6], *G.* Spadaccini[1,2], *C.* Spitaleri[7,6], and *A.* Tumino[8,6]

[1] *Dip. di Fisica "E. Pancini", Università di Napoli Federico II, I-80126 Napoli, Italy*
[2] *INFN-Sezione di Napoli, I-80126 Napoli, Italy*
[3] *Institut de Physique Nucléaire, CNRS-IN2P3, Univ. Paris-Sud, Université Paris-Saclay, 91406 Orsay Cedex, France*
[4] *INFN - Sezione di Catania, Via S. Sofia, I-95125 Catania, Italy*
[5] *Universidad Nacional Autónoma de México, A.P. 20-364, Cd.Mx, D.F. 01000 México*
[6] *INFN - Laboratori Nazionali del Sud, Via S. Sofia, I-95125 Catania, Italy*
[7] *Dip. di Fisica e Astronomia, Università di Catania, Via S. Sofia, I-95125 Catania, Italy*
[8] *Facoltà di Ingegneria ed Architettura, Università Kore, I-94100 Enna, Italy*

Abstract. The α-decay path of the Hoyle state in ^{12}C (E_x = 7.654MeV) represents one of the most challenging questions of modern nuclear physics. Its knowledge can help in the understanding of cluster configurations in light nuclei and the possible existence of Bose-Einstein condensates in nuclei. In stars, it is involved in the so-called 3α process, where the ^{12}C nucleosynthesis occurs. We studied the ^{14}N(d, α_2)^{12}C(7.654) reaction at 10.5MeV incident energy to probe its direct decay component. We found, with a precision higher of a factor 5 than any other previous experiment, an almost total absence of direct decays by-passing the ground state of ^{8}Be. From our analysis, a new upper limit of such decay is found at 0.043% (95% C.L.). Astrophysical 3α process reaction rate calculations have to be accordingly revised.

1 Introduction

The majority of the constituents of our universe are produced in stars by means of nuclear reactions, whose efficiency is regulated by the structure properties of the nuclei involved [1]. Studying nuclear reactions at low energy and the spectroscopy of nuclei is therefore important to understand how and to what extent elements are produced in stars [2–6]. A fundamental question in nuclear astrophysics is certainly how the production of carbon, one of the major constituents of the universe and human beings, could occur [7]. Such question has attracted the interest of nuclear physicists and astrophysicists since very long time. When a star leaves the main sequence, after the proton burning, it is mainly constituted by helium nuclei. In such a scenario, the way for the nucleosynthesis to proceed towards

⋆e-mail: dellaquila@na.infn.it

heavier nuclei consists in the so-called *helium burning* [1], where nuclei of helium are first transmuted into carbon and then in heavier elements via subsequent radiative captures. The first part of the process, i.e. the trasmutation of 3 α particles into a ^{12}C ($3\alpha \rightarrow ^{12}$C), is anything but trivial. This process proceeds via the initial fusion of two α particles into the ^{8}Be, followed by the radiative capture of a third α, $\alpha+^{8}\text{Be}\rightarrow^{12}\text{C}+\gamma$ [8, 9]. Anyway, due to the extremely short lifetime of the unbound ^{8}Be (formed in the intermediate stage and having a lifetime of the order of 10^{-16} s), the efficiency of the whole process is limited and, by simply assuming a non-resonant capture of the third α, it is not possible to explain the carbon to oxygen abundance ratio observed in the universe [10]. This apparent contradiction was solved by Fred Hoyle with an interesting hypothesis formulated in 1953: the radiative capture of the third α takes place via a resonant process through a $J^{\pi} = 0^{+}$ state of ^{12}C located close to the α emission threshold of ^{12}C($\approx$ 7.4 MeV) [11, 12]. The existence of this state was then confirmed, a few years later, by Cook and collaborators [13], lying at an excitation energy of 7.654MeV. This state is nowadays known as the *Hoyle state* of ^{12}C [7].

The production rate of carbon in such a *sequential* process can be calculated from the competition between radiative decays and the α-decay of the Hoyle state. This clearly depends on the structure properties of the Hoyle state, and the following empirical equation determines it [7]:

$$< \sigma v > \propto \frac{\Gamma_{\alpha}\Gamma_{rad}}{\Gamma} e^{-\frac{E_R}{k_B T}} \tag{1}$$

where E_R is the Hoyle state energy of resonance, k_B the Boltzmann constant and Γ the total level width. The latter is fully dominated by the α decay width, i.e. $\Gamma = \Gamma_{\alpha} + \Gamma_{rad} \approx \Gamma_{\alpha}$, and so it results in a dynamical equilibrium $^{4}\text{He} + ^{8}\text{Be} \leftrightarrows ^{12}\text{C}^{\star}$ only broken by the small leakage to the ^{12}C ground state given by the radiative decays.

The knowledge of the properties of the Hoyle state allows then to determine the reaction rate of the 3α process [14]. However, while in stars that burn helium at temperature grater than 10^{8} K it is dominated by the resonant process (called sequential), for lower temperatures another mechanism could be present and dominate the reaction rate: the so-called *direct* (non-resonant) process. The first of the two mechanisms is fully dominated by the Hoyle state process and by its *sequential decay* (SD) width, while in the second the 3α do not have energetically access to the intermediate resonances. Anyway, in the presence of a non-vanishing *direct decay* (DD) width of the Hoyle state, the second process may also occur through the low energy tail of the Hoyle state [15, 16], being the corresponding rate significantly enhanced with respect to the commonly adopted low energy extrapolation reported in NACRE [17]. In order to fully understand these processes, and the validity of the NACRE empirical rate formula, a precise knowledge of the decay pattern of the Hoyle state has to be experimentally achieved.

On the theoretical point of view, several refined microscopic calculations have been developed to describe the 3α process, improving the simple empirical description reported by NACRE. Ogata and co-workers [14] showed an enhancement of the 3α reaction rate at low temperatures of several orders of magnitude with respect to the NACRE assumption (26 orders of magnitude at $T = 0.01GK$), as an effect of the direct processes. Following this work, different calculations with other quantum three-body approaches have been performed [16, 18–20] but the resulting picture remained quite puzzling and far from a firm conclusion. More recently, Akahori and collaborators developed an imaginary-time formalism approach to the problem of the 3α reaction rate, producing values in good agreement with the NACRE rate [21].

In nuclear structure, the Hoyle state in ^{12}C is crucial to better understand clustering phenomena and the structure of light nuclei. Given the bosonic nature of the α-particles, the inter-cluster separation of the 3α system could be large enough that their internal structure can become no longer important. Under this peculiar condition one can expect the appearance of a Bose-Einstein condensate [22, 23],

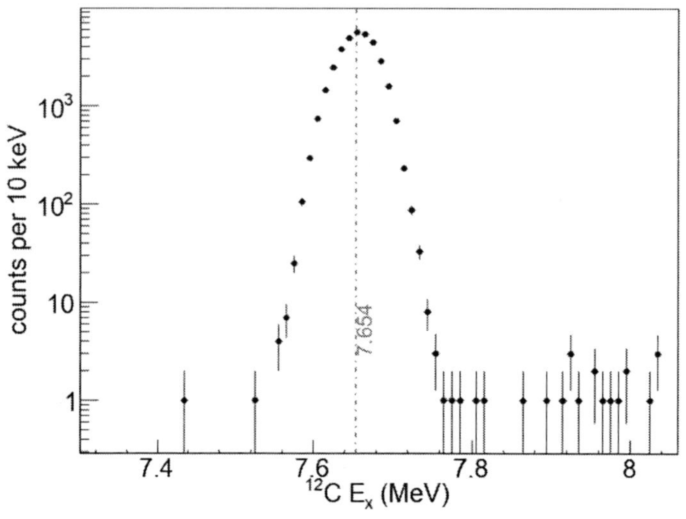

Figure 1. ^{12}C excitation energy spectrum obtained by means of invariant mass techniques on the 3α detected in the hodoscope. A line indicates the expected position of the Hoyle peak. The counts outside the peak region are associated to background events.

a very extreme case due to the limited size of the system. Such possibility could be linked to the possible presence of a non-vanishing direct decay width, as discussed in [24].

Recently, a quite large number of experiments have been performed in order to measure the direct decay (DD) width of the Hoyle state to the sequential (SD) one [24–29]. Anyway, no definitive conclusions can be drawn regarding the possible existence of a non-vanishing DD width and the precision of the state of art experiments, see [29] for example, was limited by the presence of a not negligible background level.

For these reasons, we performed a new high-precision experiment with the aim of solving the present ambiguities in the experimental knowledge of the DD width of the Hoyle state in ^{12}C with an unprecedented ultra-low background level.

2 Experimental results

Our experiment was performed at the 15 MV Tandem accelerator of INFN-Laboratori Nazionali del Sud (Catania). A highly collimated deuteron beam was accelerated at 10.5 MeV and delivered to a Melamine ($C_3N_6H_6$) target, with intensities not exceeding 4 nA to minimize possible background events due to spurious coincidences on the detectors. The ^{14}N(d,α) reaction was used to populate the Hoyle state in the residual ^{12}C. We used a two-stage telescope, placed at backward angles, and the anti-coincidence technique, see more details in Refs. [30, 31], to detect the backward recoiling α ejectile. In such a way, we are free from light ejectile contaminants coming from elastic scattering or transfer reactions induced by deuterons. The measured ejectile energy spectra, thanks to the presence of two bodies in the reaction exit channel, can be easily translated in excitation energy of the ^{12}C residual by using momentum and energy conservation laws. This apparatus allows us to tag, by means of energy cuts in the backward emitted ejectile spectrum, excitation energy and recoling direction of the residual

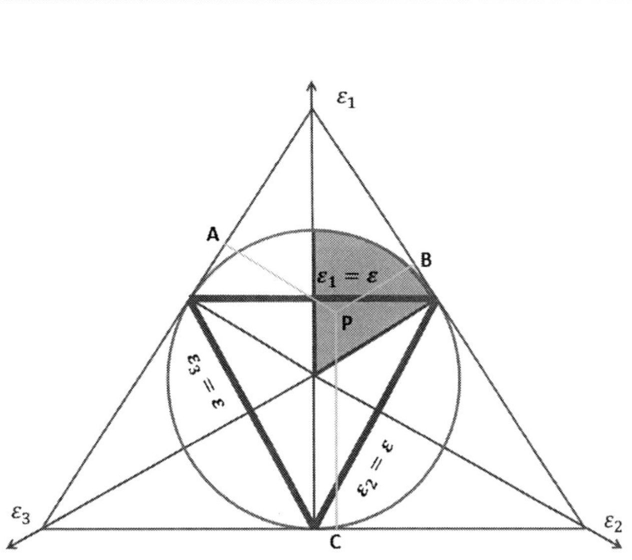

Figure 2. Representation of *normalized energy* coordinates ($\varepsilon i, j, k$) to build a symmetric Dalitz plot. By selecting $\varepsilon_i > \varepsilon_j > \varepsilon_k$ data collapse into the shadowed region. The red horizontal band indicates, schematically, the region where sequential decays should lie.

^{12}C nucleus, with extremely low background level. In kinematical coincidence with the ^{12}C*(7.654) residuals we placed an hodoscope of 64 independent silicon cells, with the aim of detecting the three α-paticles resulting from the α-decay of the Hoyle state. The telescope is constituted by 4 square blocks of 4×4 silicon detectors, each of them analogous to the second detection stage of one module of the OSCAR hodoscope (see Ref. [32] for details).

To analyze data, we restricted the events to the ones where 4 particles are detected. Such condition implements a further selectivity on the decays of the Hoyle state. As a confirmation of this, when selecting 4-particles events, the energy spectrum obtained with the backward telescope shows a strong reduction of the background under the peak associated to the Hoyle state, as shown in Ref. [33]. Invariant mass techniques, as described for example in [34], were used to unveil the decay path of the Hoyle state. Fig. 1 reports the excitation energy spectrum obtained from the invariant mass of the three particles, assuming that they are α-particles. Black points with error bars represent the experimental data and a green dashed line indicates the expected position of the Hoyle state. As clearly visible, a peak is observed centered at the correct value of energy. We obtain a precision of 1 keV in the position of the peak and the distribution has a FWHM of the order of 47 keV, indicating the high-precision level of our apparatus in invariant mass reconstruction. The contributions that lie outside of the peak region are useful to estimate the background level under the peak. It amounts at about 0.036% of the whole peak integral.

A very useful method to geometrically visualize the decay in 3α is using the so-called symmetric Dalitz plot [35]. A symmetric Dalitz plot can be constructed by using the kinetic energy of the three particles in the reference frame where the emitting nucleus is at rest. Calling $E_{i,j,k}$ these values, one can define the Dalitz plot coordinates by using the *normalized decay energies* $\varepsilon_{i,j,k} = E_{i,j,k}/(E_i + E_j + E_k)$. Since the sum of each of the coordinates is a constant, experimental points, corresponding to ($\varepsilon_i, \varepsilon_j, \varepsilon_k$) energy coordinates, are localized within an equilateral triangle (see figure 2). Moreover, being $\varepsilon_{i,j,k}$ constrained by the energy conservation law in the decay, i.e. $\varepsilon_i + \varepsilon_j + \varepsilon_k = 1$, not every combination of

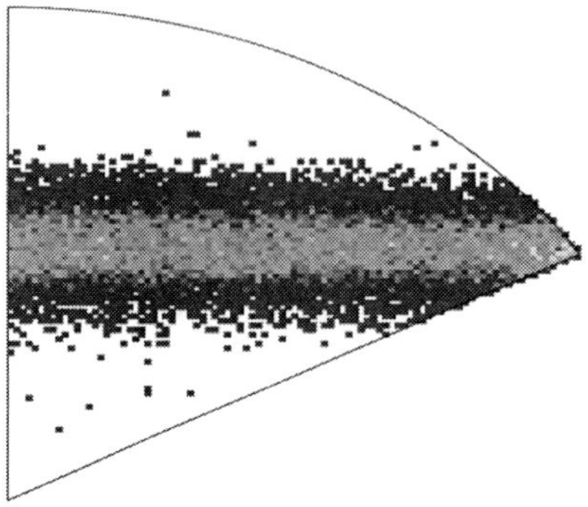

Figure 3. Experimental Dalitz plot of the 3α decay from the Hoyle state. A straight horizontal line is observed and associated to SD events. The counts outside this region are topologically compatible either with background events or signals of DD. Their amount is significantly lower than previous works.

these values is permitted and, as a result of this, data is confined within the green circle in figure 2. The distribution of experimental points in such a graph allows to distinguish different decay paths. In the case of a direct decay to the phase space, for example, energy can be shared with any of the possible combinations $\varepsilon_{i,j,k}$, and the circle is almost uniformly filled. For the case of sequential decays, instead, the energies of the decay have to obey to more restrictive constraints, resulting in the red bands shown in figure. A similar plot can be generated in cartesian coordinates by using the following equations:

$$x = \sqrt{3}(\varepsilon_j - \varepsilon_k)$$
$$y = 2\varepsilon_i - \varepsilon_j - \varepsilon_k \tag{2}$$

Finally, by selecting $\varepsilon_i > \varepsilon_j > \varepsilon_k$ the data collapse into the shaded region of Figure 2, where the red horizontal band is associated to sequential decay processes, while the remaining part of the sector will be populated by direct events.

In figure 3, we present our experimental Dalitz plot. Compared to the ones observed in the state of art experiments [27, 29], the amount of counts outside the horizontal band is much reduced. It can be associated either to DD events or to background. This confirms the low-background of our finding.

To quantitatively estimate the DD contribution we used the radial projection of the symmetric Dalitz plot. It corresponds to the highest decay energy of the three particles measured in the decay center of mass, and it is usually indicated in the literature as ε_i [29]. It is shown in the Fig. 4 from the data of the present experiment. Here, a peak at about 0.5 represents SD decays, while a distribution from about 0.33 to 0.67 is expected in the case of DD. We performed a detailed simulation to take into account the effect introduced by the experimental apparatus and the whole reaction mechanism. The result of our simulation is shown in the figure as a dashed line for the case of 100% SD. We identify a number of counts not compatible with a fully SD mechanism. Taking into account the background level, and by using the Feldman and Cousin's approach for small signals in the presence of small background levels, we found DD signals to be statistically not significant. We place a new upper limit

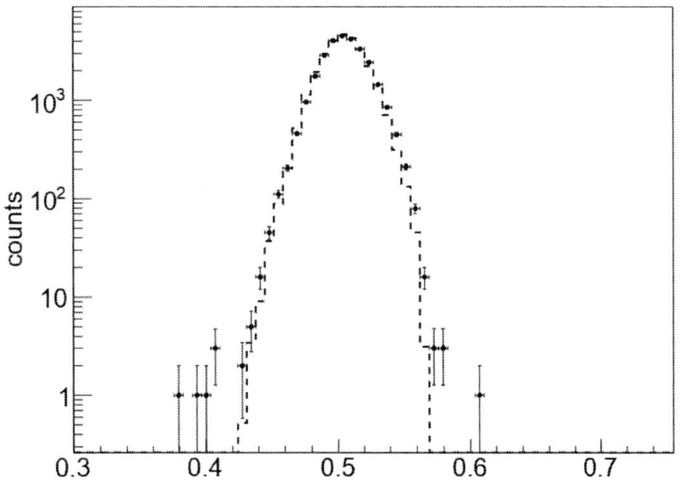

Figure 4. Distribution of the largest normalized energy ε_i in the decays of the Hoyle state. They are studied by means of invariant mass technique on experimental Hoyle state decay data measured in the hodoscope.

to DD of 0.043% with a confidence limit (C.L.) of 95% [36]. This result is about a factor 5 lower than the present state of art experiment.

3 Conclusions

The DD of the Hoyle state in ^{12}C was measured by means of a new high-precision and low background experiment. We produced ^{12}C recoils excited in their Hoyle state by means of the ^{14}N(d, α_2)^{12}C(7.654) reaction at 10.5MeV incident energy. To tag excitation energy of the ^{12}C recoiling nuclei, we used a backward angle anti-coincidence telescope. The decay of such ^{12}C is then studied via invariant mass techniques of the 3α decay, detected by means of a hodoscope detector.The use of a hodoscope made by 64 silicon independent cells allows us to avoid the background typical of the use of a single strip detector. With this experiment, we are able to place an improved upper limit for direct decay branches of the Hoyle state of 0.043% (95% C.L.). This result was recently published in [36] and has a precision of about a factor 5 better than any other previous investigations. Analogous results were accomplished by a recent experiment performed by the group of Birmingham [37]. Our results do not support a revision of the fully sequential NACRE 3α reaction rate [17] at high and intermediate temperatures ($T > 0.028$ GK). In the lower temperature region, where the 3α process is dominated by the non-resonant capture involving three α particles [21], the present results may instead give some constraints to microscopic models that assume possible contributions from 0^+ continuum states [16].

Very recently, the CHIMERA group of LNS has designed and carried out an experiment to increase the precision in the knowledge of the γ decay width of the Hoyle state. In this experiment, they benefited of a large acceptance detector [38, 39] and a high-granularity array for charged particles [40, 41]. Further attempts to better constraint the experimental knowledge of the radiative decay width of the Hoyle state, together with the present improved limit of the DD width, are particularly important in the effort of understanding clustering phenomena in light nuclei and the role of the Hoyle state in stars that burn helium at low temperatures.

References

[1] C.E. Rolfs, W.S. Rodney, *Cauldrons in the Cosmos: Nuclear Astrophysics*, 5th edn. (The University of Chicago Press, Chicago and London, 1997)
[2] C. Spitaleri et al., Phys. Rev. **C95**, 035801 (2017)
[3] I. Lombardo et al., J. Phys. G. **43**, 45109 (2016)
[4] I. Lombardo et al., Phys. Lett. B **748**, 178 (2015)
[5] I. Lombardo et al., J. Phys. G **40**, 125102 (2013)
[6] J.J. He, I. Lombardo, D. Dell'Aquila, Y. Xu, L.Y. Zhang, W.P. Liu, Chin. Phys. C **42**, 15001 (2018)
[7] M. Freer, H. Fynbo, Prog. Part. Nuc. Phys. **78**, 1 (2014)
[8] E. Opik, Proc. R. Irish Acad. A **54**, 49 (1951)
[9] E. Salpeter, Phys. Rev. **88**, 547 (1952)
[10] M. Freer, Rep. Prog. Phys. **70**, 2149 (2007)
[11] F. Hoyle et al., Phys. Rev. **92**, 1095c (1953)
[12] F. Hoyle, Astrophys. J. Suppl. Ser. **1**, 121 (1954)
[13] C. Cook, W. Fowler, C. Lauritsen, T. Lauritzen, Phys. Rev. **107**, 508 (1957)
[14] K. Ogata, M. Kan, M. Kamimura, Prog. Theor. Phys. **122**, 1055 (2009), 0905.0007
[15] K. Nomoto, F.K. Thielemann, S. Miyaji, Astron. Astrophys. **149**, 239 (1985)
[16] N.B. Nguyen, F.M. Nunes, I.J. Thompson, Phys. Rev. **C87**, 054615 (2013), 1209.4999
[17] C. Angulo et al., Nucl. Phys. A **656**, 3 (1999)
[18] E. Garrido, R. de Diego, D.V. Fedorov, A.S. Jensen, The European Physical Journal A **47**, 102 (2011)
[19] S. Ishikawa, Phys. Rev. C **87**, 055804 (2013)
[20] N.B. Nguyen, F.M. Nunes, I.J. Thompson, E.F. Brown, Phys. Rev. Lett. **109**, 141101 (2012), 1112.2136
[21] T. Akahori, Y. Funaki, K. Yabana, Phys. Rev. C **92**, 022801 (2015)
[22] Y. Funaki, H. Horiuchi, W. von Oertzen, G. Röpke, P. Schuck, A. Tohsaki, T. Yamada, Phys. Rev. C **80**, 064326 (2009)
[23] A. Tohsaki, H. Horiuchi, P. Schuck, G. Röpke, Phys. Rev. Lett. **87**, 192501 (2001)
[24] A. Raduta et al., Phys. Lett. B **705**, 65 (2011)
[25] M. Freer et al., Phys. Rev. C **49**, R1751 (1994)
[26] J. Manfredi, R.J. Charity, K. Mercurio, R. Shane, L.G. Sobotka, A.H. Wuosmaa, A. Banu, L. Trache, R.E. Tribble, Phys. Rev. C **85**, 037603 (2012)
[27] O. Kirsebom et al., Phys. Rev. Lett. **108**, 202501 (2012)
[28] T. Rana et al., Phys. Rev. C **88**, 021601(R) (2013)
[29] M. Itoh et al., Phys. Rev. Lett. **113**, 102501 (2014)
[30] W. Koenig et al., Il Nuov. Cim. **39**, 9 (1977)
[31] D. Dell'Aquila et al., J. Phys. Conf. Ser. **876**, 012006 (2017)
[32] D. Dell'Aquila et al., Nucl. Instrum. Meth. Phys. Res. A **877**, 227 (2018)
[33] Dell'Aquila, D. et al., EPJ Web Conf. **165**, 01020 (2017)
[34] D. Dell'Aquila et al., Phys. Rev. C **93**, 024611 (2016)

[35] R. Dalitz, Philos. Mag. **44**, 1068 (1953)
[36] D. Dell'Aquila et al., Phys. Rev. Lett. **119**, 132501 (2017)
[37] R. Smith, T. Kokalova, C. Wheldon, J.E. Bishop, M. Freer, N. Curtis, D.J. Parker, Phys. Rev. Lett. **119**, 132502 (2017)
[38] D. Dell'Aquila et al., EPJ Web of Conf. **117**, 06011 (2016)
[39] G. Cardella et al., Nucl. Instr. and Meth. A **799**, 64 (2015)
[40] D. Dell'Aquila, Il Nuov. Cim. C **39**, 272 (2016)
[41] E.V. Pagano et al., EPJ Web of Conf. **117**, 10008 (2016)

EPJ Web of Conferences **184**, 01006 (2018) https://doi.org/10.1051/epjconf/201818401006
9th European Summer School on Experimental Nuclear Astrophysics

Resonant Elastic Scattering

Francois de Oliveira Santos
[1]*GANIL, CEA/DSM-CNRS/IN2P3, Boulevard Henri Becquerel, Caen Cedex 5, F-14076, France*
e-mail: *oliveira@ganil.fr*

Abstract. Elastic scattering of nuclei at energies typically below 10 MeV/nucleon can be used as a powerful method for studying nuclear spectroscopy. Resonances are observed in the excitation function, corresponding to unbound states in the compound nucleus. The analysis of the shape of these resonances can provide the excitation energy, the total width, the partial width, and the spin of the excited states.

1 The Rutherford scattering

The experimental study of nuclear collisions started when Marsden, Geiger, Rutherford and their collaborators directed alpha particles emitted by a radioactive source of radium onto different metallic foils (publication in 1909 [1]). They observed that a part of the alpha projectiles interacted "strongly" with the matter and back-scattered at large angles (beyond 90 degrees). In 1911, Rutherford published [2] an interpretation of these data. In his model, the atoms are made of a nucleus positively charged and containing most of the atomic mass, surrounding by a halo of light electrons negatively charged. The positively charged alpha particles were back-scattered by the strong Coulomb repulsion of the positively charged atomic nuclei. In 1914, Darwin [3] derived a formula for the cross section of this "Rutherford elastic scattering":

$$\frac{d\sigma}{d\Omega}_{Rutherford} = \left(\frac{Z_1 Z_2 e^2}{E\sin(\frac{\theta}{2})}\right)^2 \tag{1}$$

where Z_2 is the charge of the target nucleus, E and Z_1 are the energy of the incident particle and its charge, scattered at center of mass angle θ (see Fig. 1 top left).

2 First observations of anomalies in the elastic scattering

The radioactive sources provide alpha particles with well defined energies. Different foils can be used to decelerate the particles to different lower energies. Then, it is possible to measure the excitation function, that is to say, the differential cross section as the function of the energy, and to compare it with the Rutherford scattering. According to the formula, the differential cross section should vary inversely proportional to the square of the energy. Deviations from this law were observed (Rutherford 1919 [5], Chadwick 1921 [6]). Figure 2 shows the results obtained by Chadwick for the scattering of alpha particles onto an hydrogen target, compared to the expected results using the Rutherford scattering formula. Large deviations are observed at higher energies.

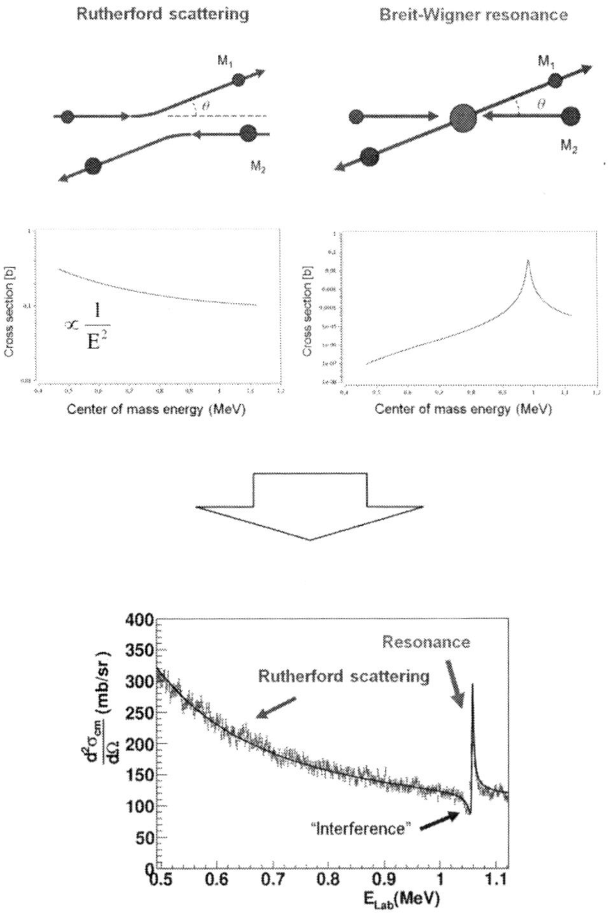

Figure 1. Top left: The Rutherford elastic scattering. The two nuclei undergo the strong Coulomb repulsion but they don't "touch" each other. The cross section can be calculated with Eq. 1. Top right: The resonant elastic scattering. The two nuclei merge and make a compound nucleus which then decays to the elastic scattering channel. The cross section can be calculated with the Breit-Wigner formula. Bottom: Excitation function of the reaction $^{14}N(p,p)^{14}N$ measured at GANIL [4] at the angle of 180° CM. The data follow very well the Rutherford elastic scattering formula, except at the energy close to $E_{lab} \simeq 1.06$ MeV where a resonance can be observed. It is interpreted as a resonance in the compound nucleus ^{15}O. As shown at energies below the maximum of the resonance, the Rutherford and the resonant contributions interfere negatively.

3 Resonances in the elastic scattering

More precise measurements made later on different targets have revealed many deviations, generally localized at certain specific incident energies. For example, Fig. 1 (bottom) shows the excitation function of the elastic scattering reaction $^{14}N(p,p)^{14}N$ measured at GANIL [4] at the angle of 180° CM. It is observed that the cross section follows very well the Rutherford elastic scattering formula, except for the energy close to $E_{lab} \simeq 1.06$ MeV where a resonance can be observed. As it is shown

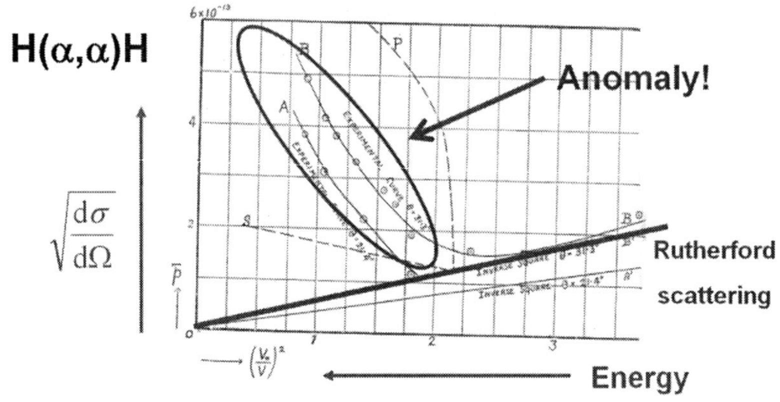

Figure 2. Square root of the differential cross section as a function of the square of the inverse of the projectile velocity, measured for the reaction H(α,α)H. In these axis, the Rutherford scattering formula is a straight line (green). Deviations from the Rutherford formula are observed at higher energies.

in Fig.3, this resonance matches perfectly to the energy of an excited state in the compound nucleus ^{15}O.

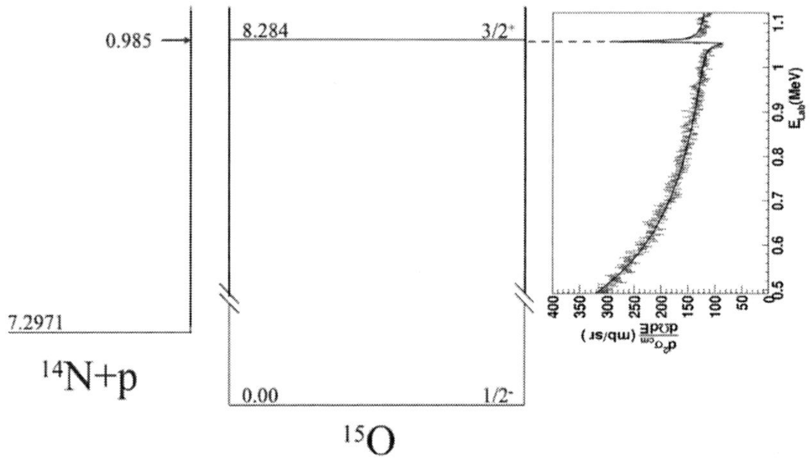

Figure 3. The energy $E_{lab} \simeq 1.06$ MeV of the resonance observed in Fig. 1, or $E_{CM} \simeq 0.985$ MeV in center of mass, above the reaction threshold ^{14}N+p (S=7.2971 MeV), matches perfectly to the energy E_x = 8.284 MeV of the 3/2$^+$ excited state in the compound nucleus ^{15}O: E_x=S+E_{CM}. The resonance is due to the existence of this excited state.

Two different reaction mechanisms can be invoked to explain the measured data. On the one hand (Fig.1, top left), the two nuclei undergo the strong Coulomb repulsion but they don't "touch" each other. On the other hand (Fig.1, top right), the two nuclei make a compound nucleus which decays to the elastic scattering channel. This reaction mechanism is called Resonant Elastic Scattering.

4 R-Matrix formalism

In reality, the two contributions, the Rutherford scattering and the resonant scattering, operate simultaneously and are indistinguishable. The final cross section is not just the sum of the two cross sections, quantum mechanics says. Amplitude of probabilities should be added, resulting sometimes into interferences between the two contributions, see Fig.1 (bottom). The R-Matrix formalism can be used to calculate, to predict or to fit the excitation function of the elastic scattering including different contributions and their interferences. The theoretical developments of this formalism would take up too much space in this article, theoretical details can be obtained in other articles, see for example Ref[7, 8]. Here, the heuristic example (without Coulomb!): the case of the s-wave scattering of a spinless neutron by a square-well potential, will be presented.

For simplicity, let's assumed the potential between the target nucleus and the projectile to be a square well with a potential depth V_0 and a radius $r = a$, see Fig. 4. The potential goes to infinity at $r = 0$, and it is zero out to $r = \infty$.

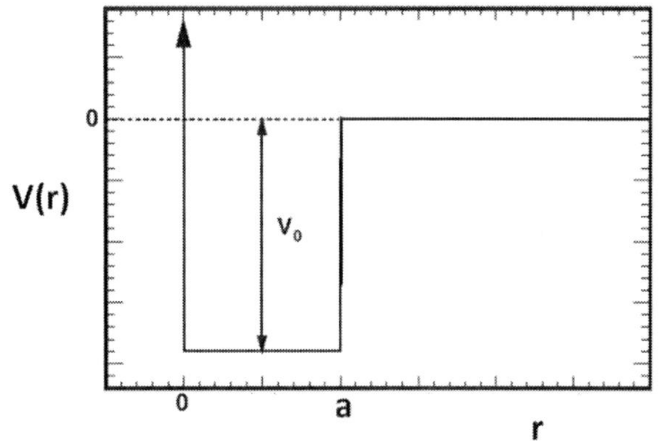

Figure 4. The potential between the target nucleus and the projectile is assumed be a square well with a potential depth V_0 and a radius $r = a$.

In this simple case, two contributions are also expected: the elastic scattering from the square-well and the resonant scattering from the compound nucleus (the neutron in the mean field of the square well). The radial part of the Schrödinger equation

$$-\frac{\hbar^2}{2m}\frac{d^2\phi(r)}{dr^2} + V_0\,\phi(r) = E\,\phi(r) \tag{2}$$

should be solved.

In the internal region ($r < a$), the solution is

$$\phi(r) = A\,\sin(K\,r) \tag{3}$$

where A is a constant and $K = \sqrt{2m(E + V_0)}/\hbar$.

In the external region ($r > a$), it is possible to show that

$$\phi(r) = B\,[e^{-ikr} - e^{2i\delta}e^{+ikr}] \tag{4}$$

and the cross section of the elastic scattering

$$\sigma(E) = 4\,\frac{\pi}{k^2}\sin^2(\delta) \tag{5}$$

where B is a constant, δ is called the phase shift, it is a function of the energy, and $k = \sqrt{2m(|E|)}/\hbar$.

The function $\phi(r)$ and its derivative should be continuous at $r = a$, this gives

$$\delta = \arctan[\frac{k}{K}\tan(Ka)] - ka \tag{6}$$

This can be rewritten

$$\delta = \arctan[k\frac{\phi(a)}{\phi(a)'}] - ka \tag{7}$$

The internal wave function can always be expanded with a complete set of basics states $X_\lambda\,(r)$

$$\phi(r) = \sum_\lambda C_\lambda\, X_\lambda\,(r) \tag{8}$$

where C_λ are different coefficients. These coefficients can be obtained with the relation

$$C_\lambda = \int_0^a X_\lambda^*(r)\,\phi(r)\,dr \tag{9}$$

The "resonant states" inside the square well are a good set of basics states. These "resonant states" can be obtained by solving the Schrödinger equation with the condition

$$\frac{dX_\lambda}{dr}(r = a) = 0 \tag{10}$$

One obtains

$$X_\lambda(r) = (\frac{2}{a})^{1/2}\,\sin(K_\lambda\, r) \tag{11}$$

with

$$K_\lambda = (\lambda + \frac{1}{2})\,\frac{\pi}{a} \tag{12}$$

At r=a, one gets

$$X_\lambda(r = a) = (\frac{2}{a})^{1/2} \tag{13}$$

Using the Schrödinger equations for $\phi(r)$ and for $X_\lambda(r)$, one can get

$$-\frac{\hbar^2}{2m}\{\frac{d^2\phi}{dr^2}X_\lambda^* - \phi\frac{d^2X_\lambda^*}{dr^2}\} = (E - E_\lambda)\phi X_\lambda^* \tag{14}$$

which gives after integration

$$-\frac{\hbar^2}{2m}\,[\,\frac{d\phi}{dr}X_\lambda^* - \phi\frac{dX_\lambda^*}{dr}]_0^a \;=\; (E - E_\lambda)\int_0^a \phi X_\lambda^*\,dr \tag{15}$$

which can be used in eq. 9, giving

$$C_\lambda = \frac{\hbar^2/2m}{E_\lambda - E}X_\lambda^*(a)\,\phi'(a) \tag{16}$$

Using this expression in eq.8, one gets

$$\frac{a\,\phi'(a)}{\phi(a)} = \frac{1}{\sum_\lambda \frac{\hbar^2/ma^2}{E_\lambda - E}} = \frac{1}{\mathrm{R}} \tag{17}$$

where R is the R-function, i.e. $\mathrm{R}=\sum_\lambda \frac{\gamma^2}{E_\lambda - E}$ with $\gamma^2 = \hbar^2/ma^2$ the reduced width. Eq. 17 can be used in eq. 7, in the approximation of a single term in λ, i.e. $\mathrm{R}=\frac{\gamma^2}{E_\lambda - E}$, giving

$$4\sin^2(\delta) = |2\sin(ka)e^{ika} - \frac{\Gamma}{(E_\lambda - E) - i\frac{\Gamma}{2}}|^2 \tag{18}$$

with $\Gamma = 2ka\,\gamma^2$ the level width, and for the cross section (eq.5)

$$\sigma(E) = \frac{\pi}{k^2}\left|2\sin(ka)e^{ika} - \frac{\Gamma}{(E_\lambda - E) - i\frac{\Gamma}{2}}\right|^2 \tag{19}$$

There are two terms in this expression. The first term corresponds to the elastic scattering onto a hard sphere (for which the potential is infinite for $r < a$ and zero for $r > a$). If the second term is neglected, when k goes to zero, $\sigma(E) \to 4\pi a^2$, which is just the geometric cross section of the square well. The second term corresponds to the Breit-Wigner resonance. The two terms may interfere since the two amplitudes are added before putting the sum to the square.

5 R-Matrix codes

There are two variants of the R-matrix formalism differing mainly by their types of applications [7]. In the calculable R-matrix, the aim is to accurately solve the Schrödinger equation mostly in the continuum and predict reaction cross sections (predict resonances properties). In the phenomenological R-matrix, the goal is to parameterize scattering data (fit of data, predict cross sections from a given set of resonance properties).

Several codes exist "in the market" to calculate the scattering cross section according to the formalism of the phenomenological R-matrix, and these are often available freely. The code "AZURE" is one of the latest and best users friendly codes. Without going into details, this code is very intuitive. The particle pairs are described in a first tab (spin, separation energy etc.). The excited states of the compound nucleus are introduced as input parameters in a second tab (excitation energies, spins, widths etc.). That's about it! More details can be obtained in the reference [9].

6 Pros and cons

Elastic scattering reactions can be used to study the structure of the compound nucleus:

- The energies E_x of the excited states can be determine from the measured energies E_R of the resonances, since $E_x = E_R + S$, with S the particle-emission threshold, see Fig. 3.
- The scattering reaction is induced most often on protons, but other particles can also be used, like alpha particles.
- Due to the interferences effects, the shape of the resonances depends on the spin of the states, see Fig. 5. Then, the analysis of the shape of the resonances can be used to assign or constrain the spins of the excited states.

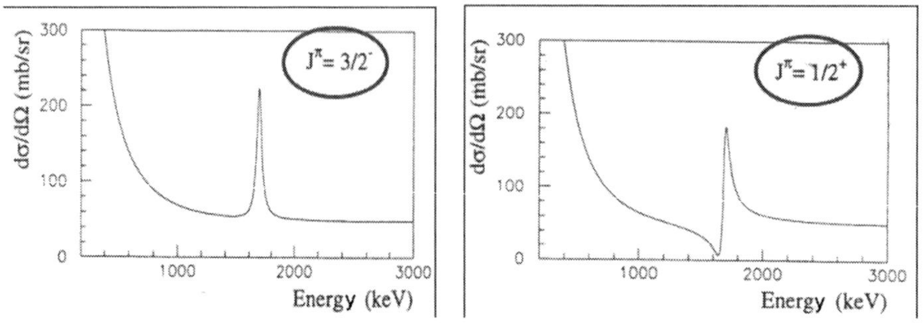

Figure 5. Elastic scattering excitation functions calculated using the same experimental conditions and exactly the same properties of the state in the compound nucleus but with different spins. The analysis of the shape of the resonance can be used to determine the spin of the states.

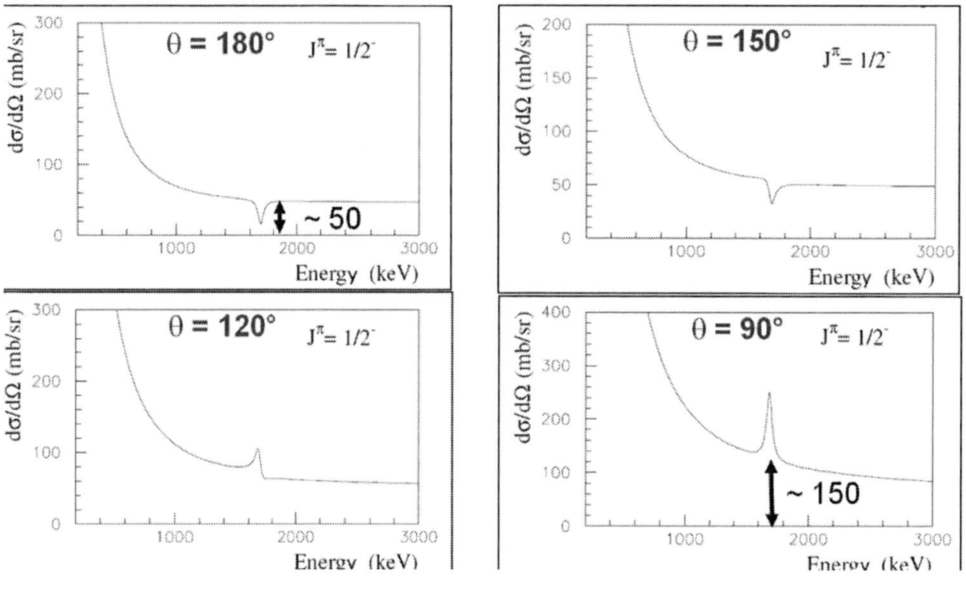

Figure 6. Elastic scattering excitation functions calculated for different angles in center of mass calculated for the reaction ^{12}C + proton (from Ref. [10]). These spectra are calculated using exactly the same properties of the resonant state in the compound nucleus. The analysis of the shape of the resonance as a function of the angle can be used to determine the spin of the state.

- The evolution of the excitation function with the angle of the scattered particles can also be used to determine the spin of the states, see Fig. 6, this is particularly useful when there are several resonances superposed.
- The width and the height of the resonances peaks can be used to measure the total and partial widths of the states.
- The differential cross sections are relatively high, typically several tens or hundreds of millibarns per steradian. Radioactive beams with relatively low intensities can be used.

- It is possible to use a radioactive beam and to measure the elastic scattering in inverse kinematics (heavy ion onto a light target nucleus). The thick target inverse kinematics (TTIK) technique (thick enough to stop the beam inside the target), first proposed by V.Z. Goldberg [11], is very well adapted to these relatively low beams intensities. The center mass energy of the interaction point (reaction vertex) can be determined from the measured energy of the protons in the lab, after correction of the energy lost in the target. The center of mass angle is related to the lab angle by $\theta_{CM} = 2\,\theta_{lab}$ (this relationship is quasi-true within the special relativity). The measured cross-section has to be corrected for different effects, mainly the fact that the effective target thickness is not constant. The energy of the scattered particles in lab is related to the center of mass energy by

$$E_{lab}(\text{scattered particle}) = 4\,\frac{m_p}{m_p + m_t}\,E_{CM}\cos(\theta_{lab}) \tag{20}$$

with m_p and m_t the projectile mass and the target mass. In the case θ_{lab}=0 and a proton target, $E_{CM} \approx \frac{1}{4}\,E_{lab}$(proton). It is possible to obtain high energy resolution spectra even with a thick target (the center of mass energy resolution is about 4 times smaller than the energy resolution in the lab), see Fig. 7.

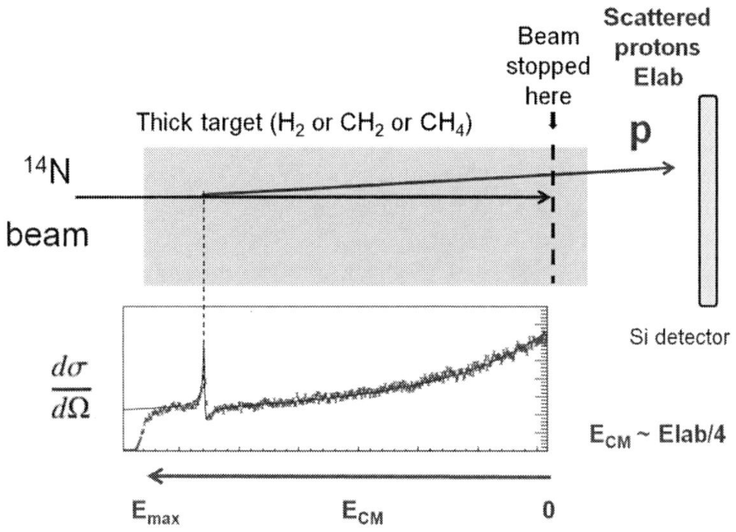

Figure 7. In the thick target technique in the case of the protons, if the energy loss of the scattered protons in the target is neglected, $E_{CM} \approx \frac{1}{4}\,E_{lab}$(proton). The center of mass energy of the reaction vertex is determined from the measured energy of the protons in the lab, after correction of the small energy losses in the target.

- The energy resolution of the measured scattered protons can be estimated using the relation: $\sigma_{Lab} = \sqrt{\sigma^2_{det} + \sigma^2_{\theta} + \sigma^2_{strag}}$, where σ_{det} is the energy resolution of the detector (σ_{det}=9 keV in the ^{14}N setting of Fig. 1), σ_{strag} is the energy straggling in the target (estimated from simulations to be lower than 5 keV in ^{14}N), and σ_θ is the energy resolution due to the aperture $d\theta$ of the detector. In inverse kinematics, it can be derived that $\sigma_\theta = \tan(\theta)E d\theta$. Therefore the degradation in energy resolution is minimal when $\theta = 0°$. For this reason, and for maximizing the ratio between the nuclear and the Coulomb contribution in the differential cross-section (see Fig. 6), the scattered protons are generally measured at very forward angles. An energy resolution of $\sigma_{Lab} \approx$ 10 keV was measured in the case of ^{14}N, which lead to $\sigma_{CM} \simeq 3$ keV in the center of mass.

The resonant elastic scattering technique has some specific limitations:

- Only states above the particle emission threshold can be studied.
- The number of counts measured in the peak is proportional to the width of the resonance. The narrower it is, the more beam time is needed to observe the peak.
- When the density of states is high, it is not possible to observe isolated resonances.
- It is not possible to observe the shape of the resonance nor the spin of the state when the resonance is narrower than the experimental energy resolution. But, if a thin target is used, it is possible to determine the cross section point by point using a beam accelerated to different energies. In principle, the energy resolution could be excellent (26 eV! in ref. [12]) provided the beam has a good energy resolution.

7 An informative experiment example

Several cases of resonant elastic scattering experiments were presented during the 2017 Santa Tecla School [13–20]. A very informative example is given in the article of Bardayan *et al* [21]. Knowledge of the astrophysical rate of the ^{18}F(p,α)^{15}O reaction is important for understanding the γ-ray emission expected from novae. The rate of this reaction is dominated at temperatures above 0.4 GK by a resonance near 7.08 MeV excitation energy in ^{19}Ne. In this study, the authors proposed to made simultaneous measurements of the ^{1}H(^{18}F,p)^{18}F and ^{1}H(^{18}F,α)^{15}O excitation functions using a radioactive ^{18}F beam at the ORNL Holifield Radioactive Ion Beam Facility. They used relatively thin target and radioactive beams accelerated to different energies. The R-Matrix analysis of the ^{1}H(^{18}F,p)^{18}F excitation function, see Fig. 8, provided many pieces of information needed to calculate the rate of the reaction. The cross section of the reaction ^{1}H(^{18}F,α)^{15}O, combined with these pieces of information, allowed to determine the spin of the state.

Acknowledgement

This publication was supported by OP RDE, MEYS Czech Republic under the project EF16-013/0001679, by LEA NuAG, by LIA COSMA and by LIA COLL-AGAIN.

References

[1] H. Geiger, E. Marsden, Proceedings of the Royal Society of London Series A **82**, 495 (1909)
[2] E. Rutherford, Phil. Mag. **6**, 21 (1911)
[3] C. Darwin, Phil. Mag. **XXVII**, 499 (1914)
[4] I. Stefan, Ph.D. thesis (Université de Caen, France, 2006)
[5] Rutherford, Phil. Mag. **XXXVII**, 537 (1919)
[6] J. Chadwick, Phil.Mag. **42, N252**, 923 (1921)
[7] P. Descouvemont, D. Baye, Reports on progress in physics **73**, 036301 (2010)
[8] C.A. Bertulani, *Nuclei in the Cosmos* (World Scientific, 2014)
[9] R. Azuma, E. Uberseder, E. Simpson, C. Brune, H. Costantini, R. De Boer, J. Görres, M. Heil, P. LeBlanc, C. Ugalde et al., Physical Review C **81**, 045805 (2010)

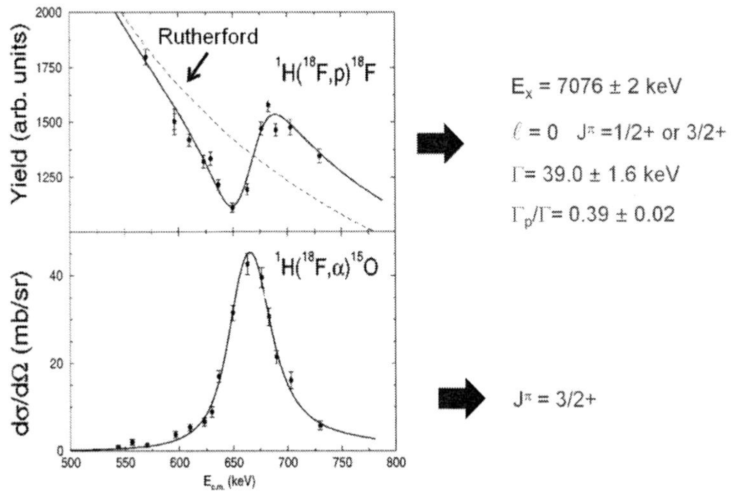

Figure 8. The reactions ^{1}H(^{18}F,p)^{18}F and ^{1}H(^{18}F,α)^{15}O were measured in the same experiment simultaneously [21]. The R-matrix formalism was used to fit the two spectra. Important spectroscopic information were extracted.

[10] Lynda Achouri (Ph. D Thesis. Universite de Caen, 2001)
[11] V. Gol'Dberg, A. Pakhomov, Physics of Atomic Nuclei **56**, 1167 (1993)
[12] S. Wüstenbecker, H. Becker, H. Ebbing, W. Schulte, M. Berheide, M. Buschmann, C. Rolfs, G. Mitchell, J. Schweitzer, Zeitschrift für Physik A Hadrons and nuclei **344**, 205 (1992)
[13] D. Torresi, L. Cosentino, P. Descouveont, A. Di Pietro, C. Ducoin, P. Figuera, M. Fisichella, M. Lattuada, C. Maiolino, A. Musumarra et al., *8Li+ α resonant elastic scattering: a tool to study cluster states in 12B*, in *Journal of Physics: Conference Series* (IOP Publishing, 2014), Vol. 569, p. 012024
[14] H. Yamaguchi, D. Kahl, Y. Wakabayashi, S. Kubono, T. Hashimoto, S. Hayakawa, T. Kawabata, N. Iwasa, T. Teranishi, Y. Kwon et al., Physical Review C **87**, 034303 (2013)
[15] F. de Oliveira Santos, P. Himpe, M. Lewitowicz, I. Stefan, N. Smirnova, N. Achouri, J. Angélique, C. Angulo, L. Axelsson, D. Baiborodin et al., The European Physical Journal A-Hadrons and Nuclei **24**, 237 (2005)
[16] M. Freer, E. Casarejos, L. Achouri, C. Angulo, N. Ashwood, N. Curtis, P. Demaret, C. Harlin, B. Laurent, M. Milin et al., Physical review letters **96**, 042501 (2006)
[17] L. Axelsson, M.J.G. Borge, S. Fayans, V.Z. Goldberg, S. Grévy, D. Guillemaud-Mueller, B. Jonson, K.M. Källman, T. Lönnroth, M. Lewitowicz et al., Phys. Rev. C **54**, R1511 (1996)
[18] M. Assie, F. de Oliveira Santos, T. Davinson, F. De Grancey, L. Achouri, J. Alcántara-Núñez, T. Al Kalanee, J.C. Angélique, C. Borcea, R. Borcea et al., Physics Letters B **712**, 198 (2012)
[19] I. Stefan, F. de Oliveira Santos, O. Sorlin, T. Davinson, M. Lewitowicz, G. Dumitru, J. Angélique, M. Angélique, E. Berthoumieux, C. Borcea et al., Physical Review C **90**, 014307 (2014)
[20] F. De Grancey, A. Mercenne, F. de Oliveira Santos, T. Davinson, O. Sorlin, J. Angélique, M. Assié, E. Berthoumieux, R. Borcea, A. Buta et al., Physics Letters B **758**, 26 (2016)
[21] D. Bardayan *et al.*, Phys. Rev. C **63**, 065802 (2001)

EPJ Web of Conferences **184**, 01007 (2018) https://doi.org/10.1051/epjconf/201818401007
9th European Summer School on Experimental Nuclear Astrophysics

Heavy Element Nucleosynthesis

Mounib F. El Eid[1]

[1]*American University of Beirut, Department of Physics, Beirut, Lebanon*

Abstract. This contribution deals with the important subject of the nucleosynthesis of heavy elements in the Galaxy. After an overview of several observational features, the physical processes responsible mainly for the formation of heavy elements will be described and linked to possible stellar sites and to galactic chemical evolution. In particular, we focus on the neutron-capture processes, namely the s-process (slow neutron capture) and the r-process (rapid neutron capture) and discuss some problems in connection with their sites and their outcome. The aim is to give a brief overview on the exciting subject of the heavy element nucleosynthesis in the Galaxy, emphasizing its importance to trace the galactic chemical evolution and illustrating the challenge of this subject.

1 Introduction

Elements formation and their evolution is linked to stellar evolution. Fusion reactions can synthesize elements up to Iron, which is the branching point between fusion and fission. Beyond iron, the synthesis of elements is possible only through neutron capture reactions, since neutrons do not experience a coulomb barrier. The main neutron-capture processes are known to be the slow s-process and the rapid r-process. Both of them occur during the late evolutionary phases of stars.

A convenient way of studying the heavy element synthesis is to see their traces in metal-poor halo stars. Their chemical compositions indicates the types of syntheses that happen in the early galaxy. This is a link to "near-field cosmology", or low-red shift universe. The high red-shift universe is not subject directly to observations. But fortunately, we have the primordial nucleosynthesis, the microwave back ground radiation and the Hubble expansion, which help us to draw conclusion about the early universe.

2 Observations

There exist now detailed abundance distributions for very metal-poor stars. The presence of heavy elements in these stars indicate the role of a distinct generation of massive stars (second generation?) stars, which were able to synthesize the observed heavy elements. The discovery of radioactive elements (Th, U) in low-metallicity stars helped to date the oldest stars in the Galaxy. In the following, several interesting observations are presented to motivate the theoretical effort explained below. The data in Fig. 1 represent a powerful tool to

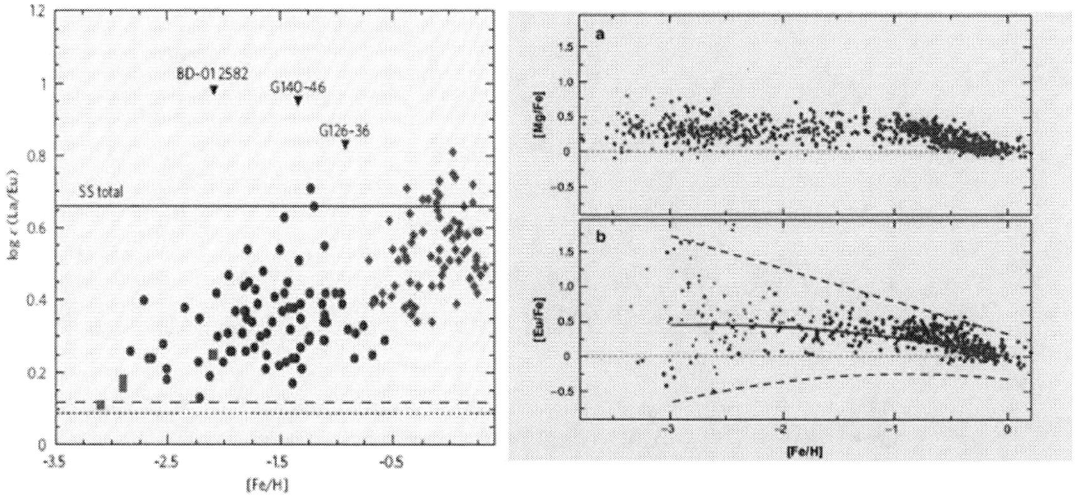

Figure 1. Left panel: Variation of abundances as a function of metallicity. These Figures are taken from [15]. Right panel: Abundances of Mg and Eu compared to Fe as a function of [Fe/H]. Note that quantities like [Fe/H] denote a logarithmic scale, for example [Fe/H]=log(Fe/H)$_{star}$-log(Fe/H)$_{sun}$. See text for details.

study the chemical evolution of the Galaxy. The left panel of this Figure shows interesting features. The abundance ratio of the Lanthanum (La, being mainly produced by the s-process nucleosynthesis) to Europium (Eu, being mainly produced by the r-process) is shown as a function of metallicity [Fe/H] for a variety of halo and disk stars. Note that line drawn near to log $\epsilon(La/Eu) = 0.1$ consistent with a pure r-process pollution, and matches very few metal-poor stars. It is also seen that the log $\epsilon(La/Eu)$ increases with increasing metallicity, where a significant contribution by the s-process starts already at [Fe/H]$\simeq -2.0$. It is still not entirely clear how the s-process became effective at such low metallicity, and which is the role of metal-poor massive stars?. A firm point is a clear presence of neutron-capture (briefly n-capture) elements in the atmospheres of metal-poor star.

The comparison between r-process rich ([Eu/Fe]$\geq$ 1.0) and r-process poor ([Eu/Fe]$<$ 1.0) stars indicates that the abundances of the n-capture elements with Z$\geq$ 56 (barium and above) are consistent with scaled solar r-process distribution. This is now termed "main r-process". In contrast, the abundances of the light n-capture species (first r-process peak around the element Sr) do not match the solar system pattern, and this is termed "weak r-process", because relatively low neutron density of about 10^{20} neutron/cm^3 is needed to produce theses species (see [?]). The right panel of Fig. 1 shows the abundances of the alpha-element Mg and those of the r-process element (Eu). The abundances of Mg have little scatter and no significant drift at low metallicity. This may imply a significant production of Mg during the early phase of the Galaxy contributed by supernova explosions of massive stars, [17]. The downward decrease of [Mg/Fe] to the solar value indicates the increase in the iron production by type Ia supernovae (exploding white dwarfs). In contrast, the data for [Eu/Fe] exhibit deviations from [Fe/H]=0 at low metallicity. A lager scatter is found as well. The scatter is attributed (see [15] and references therein) to an early unmixed galaxy, where the supernovae were widely scattered. Moreover, the data suggest that the r-process production was rare in

Table 1. Summary of some characteristics of the s-process and r-process

	s-process	r-process
Neutron density [cm^{-3}]	$10^7 - 10^{11}$	$10^{20} - 10^{30}$
Astrophysical Site	AGB Stars, Massive Stars	Supernova type II Neutron star mergers
Abundances correlated with	neutron capture cross section	beta-decay lifetimes, n-separatio
Neutron source	$^{13}C(\alpha,n)^{16}O$, $^{22}Ne(\alpha,n)^{25}Mg$	begin with neutrons & protons
Temperature range	T= $(1.0-3.5)\times 10^8$ K, T$\geq 10^9$ K	T=$(1-10)\times 10^9$ K
Basic Characteristic	Secondary process	Primary process

the early Galaxy, and only few stars produced elements such as Eu. It seems that in thye earely galaxy, there was no strong coupling between the production of Mg and Fe and Eu. Note that the downward decrease of [Eu/Fe] to the solar value is again due to the increasing iron production by type Ia supernovae.

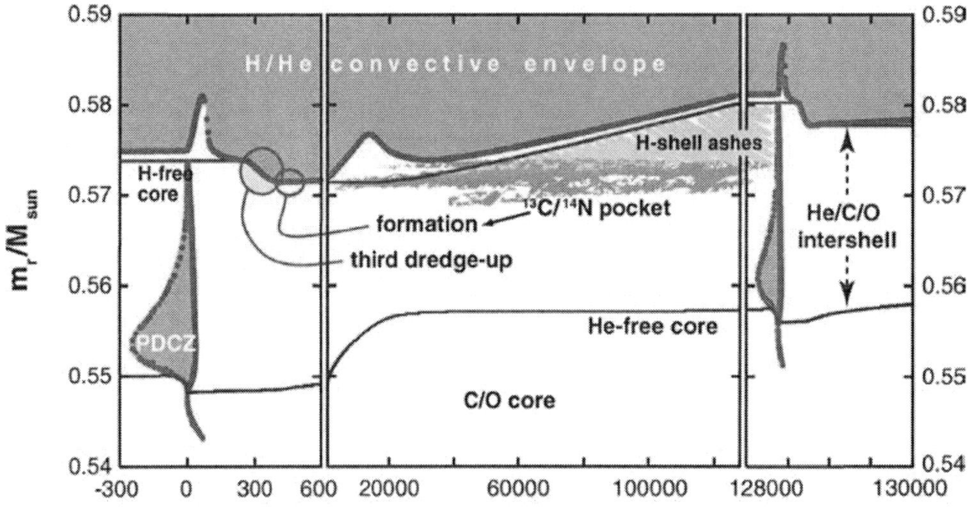

Figure 2. Structure of an AGB star during thermally pulsating phase according to [8] in case of a 2 $M_\odot$ star of initial solar-like composition. The green (or darker) areas are convective zones. See text for details

3 Basic comments on the s-process and r-process

In Table 1, a comparison between the s-process and the r-process is given. in case of the s-process, a temperature range of $(1.0-3.5)\times 10^8$ K is needed in core He-burning of massive stars (M$\geq$ 15 M$_\odot$), or during the thermal pulses of AGB stars, while T$\geq 10^9$ K is needed during shell carbon-burning in massive stars ([6]). The temperature range for the r-process is higher whether in neutron star mergers, or in an exploding supernova, where it starts near the edge of the newly formed neutron star. In this scenario, the iron-group elements are synthesized in situ as explained below. Regarding the neutron sources, those for the s-process are: $^{13}C(\alpha,n)^{16}O$ in AGB stars, while $^{22}Ne(\alpha,n)^{25}Mg$ becomes effective in massive stars, and could be activated in massive AGB stars (see [8] concerning this point)

3.1 Comments on the s-process

Basically, the s-process is a flow concept process controlled by slow neutron-capture reactions with neutron density in the range $10^7 - 10^{11}$ neutrons/cm^3. The produced neutrons thermalize very quickly by s-wave scattering and can be described by Maxwell-Boltzmann distribution. Due to the relatively low neutron density, any unstable nucleus created by n-capture will undergo β^--decay. In other words, the condition $\tau_\beta \ll \tau_{n\gamma}$,where the τ's denote lifetimes, is fulfilled meaning that beta-decay dominates the flow, so the s-process path remain close to the valley of beta-stability (for example see Fig. 9.13 in the text by [14]). However, if the neutron density is high enough like, 10^{11} neutrons/cm^3, then branching at an unstable nucleus is possible. An example dealing with Zr-isotopes is given in [8].

3.2 Comments on the r-process

The r-process is also based on a flow concept (a review is given by [5]). But in contrast with the s-process, it is a primary process in the Galaxy, since it creates its own seed nuclei. A basic feature of this process is the very high neutron density as seen in table 1. Consequently, the condition $\tau_{n\gamma} \ll \tau_\beta$ is fulfilled in this case, and the neutron capture reactions dominate the flow but not at all the time. Due to the fast addition of neutrons, neutron-rich nuclei are formed far from the valley of beta-stability, so that the neutron binding energy decreases and tends to approach the neutron drip line. For a given temperature T and neutron density an equilibrium between the (n,γ) and (γ,n) reactions becomes possible, which is called *waiting points* for each isotopic chain with proton number Z. This behavior is seen in the Fig. 9.13 in the text by [14]. In particular, the neutron magic nucleus ^{130}Cd with N=82 is a special waiting point for the r-process with a β^--lifetime of (127 ± 2) ms according to the RIKEN-results [12]. This means the abundances resulting from the r-process are correlated with the beta decay life times and neutron separation energy in contrast with those from s-process which are correlated with the neutron capture cross section.

4 Stellar Sites

The stellar sites of the heavy element synthesis are not easy to describe. In the following we describe briefly possible sites of the s-process and the r-process. No attempt is made to describe the secondary p-process producing proton-rich nuclei bypassed by the s-process (see review by [13].

4.1 s-process stellar site

As mentioned in Sect. 2, the s-process started later in the Galaxy, because it is a secondary process which needs iron seed nuclei. The main component of this process comprising nuclei with mass numbers A$\geq$ 90 is produced during the thermally pulsing AGB (TP-AGB) stars in the mass range (1-8) $M_\odot$. The dominant contribution matching the solar pattern distribution comes from mass range roughly (1-3) $M_\odot$ ([3]).

Careful modeling of TP-AGBs is required, in particular to understand the formation of the carbon-rich stars and the production of the main neutron source for the s-process $^{13}C(\alpha,n)^{16}O$. This is highly dependent on the mixing of protons from the convective envelope into the helium and carbon rich layers during a thermal pulse. A useful review on general properties of AGB stars is given by ([8], [16], [9]). Fig. 2 shows two successive thermal pulse during the AGB phase of a 2 $M_\odot$ star. The crucial issue in such calculations is the treatment of the so called third dredge up (TDU). On ground of the usual treatment of convection instability according to the well-known Schwarzschild criterion, the extension of the H-rich convective envelope after the He-shell flash (or thermal pulse) is not deep enough in order to inject protons into the carbon enriched layers. This is requited to activate the neutron source for the s-process nucleosynthesis, which is created by the reaction $^{12}C(p,\gamma)^{13}N(\beta^{+})^{13}C(\alpha,n)^{16}O$. A key problem is to get the sufficient amount of ^{13}C in order to explain the s-process abundances in the Galaxy ([19]). In [19], an important point is emphasized that the understanding of the s-process should not be restricted to matching the solar s-process abundances, but should concern with aspects of the chemical evolution in the Galaxy as well.

The production of ^{13}C is linked tightly to the to the TDU. Several effects play a role in fixing the extension of the TDUP, like extra diffusion of protons termed "overshooting" (see discussion in ([8]), [2], [4]), rotationally-induced mixing ([11]), or a recent attempt considering the effect of mixing by magnetic buoyancy in the framework of magnetohydrodynamics ([20]). More work is needed to better understand mixing processes during the AGB phases of stars.

4.2 r-process stellar site

The favored site of the r-process nucleosynthesis was the high entropy wind (HEW) in core collapse supernovae (CCSN). A basic scenario is ([10], [21]) that the streaming out neutrinos from the forming neutron star represent a neutrino-driven wind starting near the surface of the newly forming neutron star with a flow of neutrons and protons. In a temperature range T=(6-10)x10^9 K, the neutrons and protons combine to form alpha-particles, but with excess of neutrons. When the temperature drops to the range T=(3-6)x10^9 K, nuclei in the iron group are able to form, which serve as seed nuclei for the r-process. This is called *alpha freeze out*. Upon further cooling to the range T=(1-3)x10^9 K, the r-process becomes effective.

However the role of neutron star mergers has changed the picture significantly (see reviews by [18], [1]). It seems now that neutron star mergers constitute the major cite of the heavy element production by the r-process, and the only source of elements beyond lead and bismuth [18]. The remarkable insight was that the neutron star mergers were identified with short duration of Gamma-ray Bursts via their IR afterglow, to be explained by the opacities of heavy element beyond the iron group.

Does this discussion mean that the core collapse supernovae (CCSN) are removed from being the site of any r-process?. An answer to this question is not straightforward. For example, the behavior of [Eu/Fe] shown in Fig. 1 with its large scatter at low metallicities may reveal the role of early CCSN of massive stars. It seems that the large scatter of [Eu/Fe] indicates production with low event rate with high amount of r-process to explain the solar abundance, and a lack of mixing into the interstellar medium of these products steaming from CCSN. In short, the role of CCSN contributing to the r-process cannot be excluded especially at low metallicities.

The signature of the r-process is important in halo stars , since they are the oldest objects in the Galaxy, providing important insight about its early evolution. A well-studied halo star is *CS22892-052* by [15] having [Fe/H]=-3.0. A striking finding is the agreement between the abundances of this star and the solar r-process abundances for $Z \geq 56$ (that is starting from Ba). But no agreement for range Z=40-50 (first r-process peak) was found. It is beyond the scope of this contribution to discuss this issue. It seems that the r-process was not happening the same way all the time as it has been thought. There is an incomplete (or weak) component similar to the s-process. More work is needed to understand the nature site of this component.

5 Brief Summary

The heavy elements and their isotopes are mainly produced by the s-process and the r-process and to less extent by the p-process. They trace back the chemical evolution of the galaxy, thus they represent a link to the low red-shift universe. Many details of these processes are still not understood, especially the physical conditions under which they operate. In case of the s-process the treatment of mixing on the AGB phase is rather challenging and needs further elaboration. Eventually three dimensional simulations are desired to at least constraint the one-dimensional description. In case of the r-process occurring in core collapse supernovae, the modeling of the explosion is still challenging, especially in light of the newest observations of the radioactive element ^{44}Ti in the supernova remnant *Cassiopeia A* as observed by the NuStar team ([7]. The site of the r-process has received a new insight after the discovery of neutron star merger GW180817 by the LIGO and VIRGO collaboration.
Acknowledgment: The author thanks the referee for careful reading and useful comments.

References

[1] Baiotti, L., Rezzola, LK., Rep. Prog. Phys. **80**, 096901 (2017)
[2] Bisterzo, S., J. Phys. Conf. Ser., 665, 012023 (2016)
[3] Busso, M. et al., Ann. rev. A&A **37**, 239 (1999)
[4] Cristallo, R. et al., APJ, 833, 181 (2016)
[5] Cowan, J.J., Thielemann, F-K., Phys. Today **57**, 47 (2004)
[6] El Eid, M.F., et al. Space Sc. Rev. **147**, 1 (2009)
[7] Grefenstette, R.W. et al., Nature, **506**, 339 (2014)
[8] Herwig, F., Ann. rev. A&A **43**, 435 (2005)
[9] Karalkas, A., Lattanzio, J.c., PASA 31, e030 (2014)
[10] Kratz, K.L., AIP Cof. Proc., **1065**, 109 (2015)
[11] Langer, N. et al. A&A **346**, 37L (1999)
[12] Lorusso, G. et al., Phys. Rev. Lett., **114**, 192501 (2015)
[13] Meyer, B.S., Ann. rev. A&A **32**, 152 (1994)
[14] Rolfs, C., Rodney, W., **Cauldrons in the Cosmos**,(Chicago University press, Chicago, 1988) p. 472
[15] Sneden, C. , et al., Ann. rev. A&A **46**, 241 (2008)
[16] Straniero, O. et al., Nucl. Phys. A, 777, 311 (2006)
[17] Thielemann, F.-K et al. , Mem. Soc. Aston. Ital. **71**, 481 (2000)
[18] Thielemann, F.K. et al., Ann. Rev. Nucl. Part. Sci., **67**, 253 (2017)
[19] Trippella, O., et al., ApJ **787**, 41 (2014)

[20] Trippella, O., et al., ApJ **818**, 125 (2016)
[21] Woosley, S. Janka, T., Nature phys, **1**, 147 (2005)

Trojan Horse Method experiments with radioactive ion beams

Marisa Gulino[1,2,⋆], *Silvio* Cherubini[1,3,⋆⋆], *Giuseppe Gabriele* Rapisarda[1], *Shigeru* Kubono[4], *Livio* Lamia[1,3], *Marco* La Cognata[1], *Rosario Gianluca* Pizzone[1], *Hidetoshi* Yamaguchi[5], *Seya* Hayakawa[5], *Yasuo* Wakabayashi[5], *Naohito* Iwasa[6], *Seigo* Kato[7], *Tetsuro* Komatsubara[8], *Takashi* Teranishi[9], *Alain* Coc[10], *Nicolas* de Séréville[11], *Fairouz* Hammache[11], *Gabor* Kiss[12], *Shawn* Bishop[13], *Dam Nguyen* Binh[5,14], *Brian* Roeder[15], *Livius* Trache[16], *Robert* Tribble[15,17], *Roberta* Spartà[1], *Iolanda* Indelicato[1], and *Claudio* Spitaleri[1,3]

[1] *INFN-LNS, Catania, Italy*
[2] *Università di Enna Kore, Enna, Italy*
[3] *Dipartimento di Fisica e Astronomia, Università di Catania, Catania, Italy*
[4] *RIKEN Nishina Center, 2-1 Hirosawa, Wako, Saitama 351-0198, Japan*
[5] *CNS, University of Tokyo, Wako Branch, 2-1 Hirosawa, Wako, Saitama 351-0198, Japan*
[6] *Department of Physics, Tohoku University, 6-6 Aoba, Sendai, Miyagi 980-8578, Japan*
[7] *Department of Physics, Yamagata University, Yamagata 990-8560, Japan*
[8] *Rare Isotope Science Project, IBS, Yuseong-daero, Yuseong-gu, Daejeon 305-811, Korea*
[9] *Department of Physics, Kyushu University, Fukuoka 812-8581, Japan*
[10] *Centre de Spectrométrie Nucléaire et de Spectrométrie de Masse, IN2P3, F-91405 Orsay, France*
[11] *Institut de Physique Nucléaire, IN2P3, F-91405 Orsay, France*
[12] *Institute for Nuclear Research (MTA-ATOMKI), Debrecen, Hungary*
[13] *TUM, Garching, Germany*
[14] *30 MeV Cyclotron Center, Tran Hung Dao Hospital, Hoan Kiem District, Hanoi, Vietnam*
[15] *Texas A& M Univ, Inst Cyclotron, College Stn, TX 77843 USA*
[16] *IFIN HH, Bucharest, Romania*
[17] *Brookhaven Natl Lab, Upton, NY 11973 USA*

Abstract. The Trojan Horse Method (THM) is an indirect method that allows to get information about a two body reaction cross-section even at very low energy, avoiding the suppression effects due to the presence of the Coulomb barrier. The method requires a very accurate measurement of a three body reaction in order to reconstruct the whole kinematics and discriminate among different reaction mechanisms that can populate the same final state. These requirements hardly match with the typical low intensity and large divergence of radioactive ion beams (RIBs), and experimental improvements are mandatory for the applicability of the method. The first reaction induced by a radioactive ion beam studied by applying the THM was the $^{18}F(p,\alpha)^{15}O$. Two experiments were performed in two different laboratories and using different experimental set-ups. The two experiments will be discussed and some results will be presented.

⋆e-mail: gulino@lns.infn.it
⋆⋆e-mail: cherubini@lns.infn.it

1 Introduction

The Trojan Horse Method (THM) has been recently extended to reactions induced by radioactive ion beams (RIB).
The method allows to get information about a nuclear reaction cross-section having two body in the final state, $A + x \rightarrow b + c$, by studying an appropriate reaction with three body in the final state, $A + a \rightarrow b + c + s$. The nucleus a, called Trojan Horse nucleus, must have a strong cluster structure x-s. So, it is possible to have events in which the nucleus a breaks-up inside the nuclear field of the nucleus A, the cluster x interacts with the nucleus A through the $A + x \rightarrow b + c$ reaction, while the cluster s acts as a spectator. If this is the case, the cross section of the $A + a \rightarrow b + c + s$ reaction can be easily factorized and used to deduce that one of the $A + x \rightarrow b + c$ reaction [1, 2].
In a THM experiment, once the Trojan Horse nucleus a has been chosen, the energy and the angular position of at least two out of the three ejected particles in the $A + a \rightarrow b + c + s$ reaction must be detected. So, making an hypothesis about the nature of the undetected nucleus, it is possible to reconstruct the whole kinematics [5] and to select the reaction channel of interest. Moreover, to apply the factorization of the cross section of the three body reaction so inferring information about the reaction of interest, only the events proceeding through a quasi free reaction mechanism must be considered.
In order to proceed through these data analysis steps, some features should be taken into account. First of all, the kinematic is very sensitive to the angles of the tracks of the outgoing products. For this reason, in the usual THM experiments, the beam spot diameter is reduced to 1 or 2 mm, the thickness of the target is kept as thin as possible, and position sensitive detectors having a resolution of the order of millimetre are used in the experimental set-up. Moreover, as the quasi-free events are a small fraction of the whole statistics (generally of the order of 10%, depending strongly on the reaction), it is mandatory to collect enough events to obtain reasonable statistical errors.
All these requirements hardly match with the typical low intensity, presence of contaminants and large divergence of RIBs. To overcome these difficulties, more efficient experimental set-up must be used when RIBs are involved in THM experiments.

2 The ^{18}F(p,α)^{15}O reaction

The first reaction induced by a RIB that has been studied by applying the THM is the ^{18}F(p,α)^{15}O. This reaction is important to understand the novae explosion phenomena. The γ-ray emission following the novae explosion is dominated by the 511 keV energy line, coming from the annihilation of positrons produced by the decay of radioactive nuclei. Among them, the ^{18}F is expected to be abundant in the novae environment, and its lifetime well matchs the timescale for novae ejects to become transparent to γ-ray emission. It is then important to know the rate of the nuclear reactions producing and destroying ^{18}F.
At the relevant temperatures, the cross section of the ^{18}F(p,α)^{15}O reaction that destroys ^{18}F is dominated by the presence of several levels in the ^{19}Ne compound nucleus around the ^{18}F-p threshold. For this reason, it is especially important to have a good reconstruction of the ^{18}F-p relative energy. Moreover, the tails of some subthreshold levels are expected to contribute to the resonance cross section, and interference effects between excited states having the same spin can also occur.
To infer information on the ^{18}F(p,α)^{15}O reaction in the low interesting energy region, the three body d(^{18}F,α^{15}O)n reaction was studied by using a ^{18}F beam impinging on a CD_2 solid target. Two different experiments were performed in two different laboratories and using different experimental set-up.
The first experiment was performed at the CRIB facility in RIKEN, Japan [3]. The second one was performed using the MARS spectrometer installed at TAMU, USA [4].

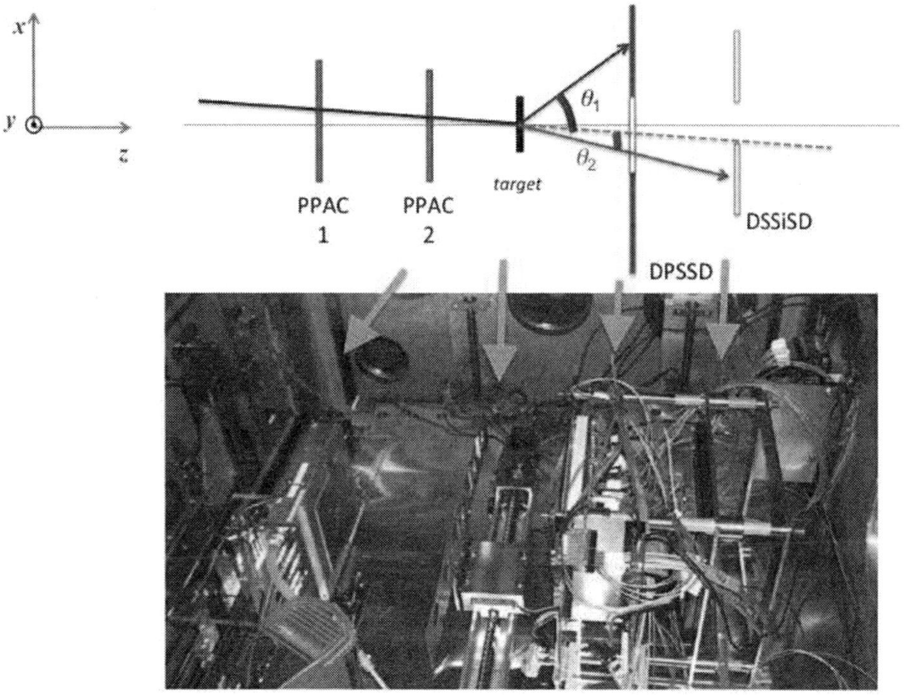

Figure 1. Scheme and photo of the experimental set-up used for the ^{18}F+d experiment performed at CRIB, RIKEN, Japan (upper view). The beam line is in the z axis; PPAC 1 and PPAC 2 track the effective beam trajector event by event; DPSSD is the plane containing 8 double position sensitive Si detectors by Hamamatsu; DSSiSD is a couple of double sided Si strip detectors. Note that the angles of the ejected particles are reconstructed with respect to the measured beam track event by event.

In order to partially overcome the problems arising from the typical low intensity, presence of contaminants and large divergence of RIBs, the experimental set-up used in both the experiments are very different from those used in standard THM experiments. Indeed, detectors covering a large solid angle were used to measure the energy and the angle of the ejected particles. In particular, in the CRIB experiment a pair of a double sided multistrip detectors (DSSiSD), manufactured by Micron Semiconductor (16 strips per side, width 3 *mm*) and placed at 287 *mm* from the target, were used to detect the heaviest ejectile. The other ejectile was detected by using the ASTRHO (A Silicon Array for TRojan HOrse) modular system of INFN Laboratori Nazionali del Sud. In the configuration used for this experiment, ASTRHO was equipped with 8 bidimensional position sensitive detectors (45×45 mm^2, spatial resolution 1 *mm*), manufactured by Hamamatsu Photonics and placed at 150 *mm* from the target. For each detector the x and y position and the energy of the impinging particles were recorded together with the time of flight. A sketch of the experimental set-up is shown in fig. 1.
The set-up used in the experiment performed at TAMU consists of two position sensitive detectors (PSDs, type X1,16 strips each), placed symmetrically at 340 *mm* from target and used to detect the heaviest ejectile, and the TECSA array[6], made up of 8 YY1-300 Micron detectors (each one with 16 arch strips), placed at 190 *mm* from target used. In fig. 2 a schematic view of the experimental

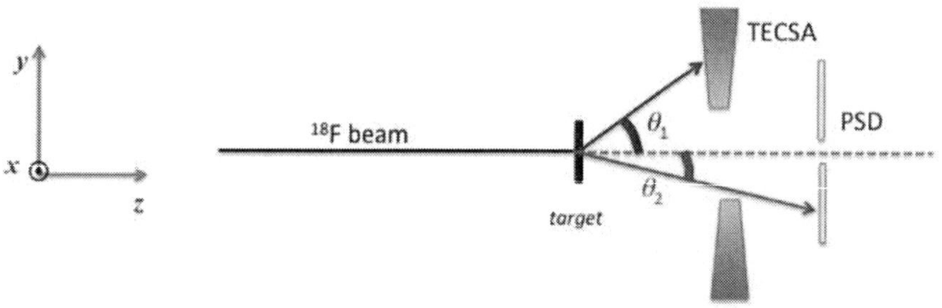

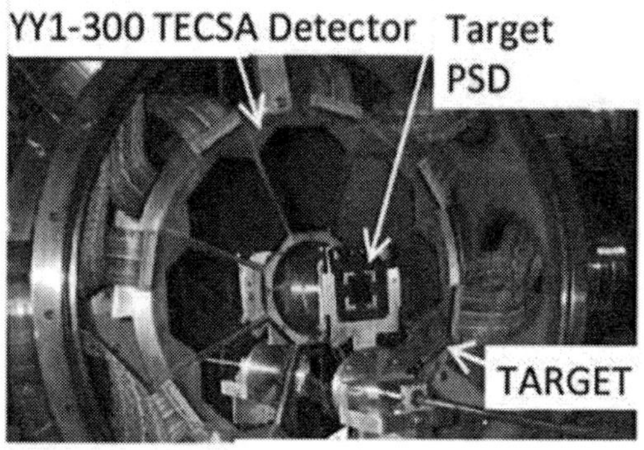

Figure 2. Scheme and photo of the experimental set-up used for the ^{18}F+d experiment performed at Mars, TAMU, USA. The beam line is the z axis; TECSA is the Si strip detector plane containing 8 annular sectors; PSD is a couple of position sensitive Si detectors.

set-up is shown.

The main difference between the two experimental set-ups comes from the use of a beam tracker in the CRIB experiment consisting of two PPAC detectors placed at 510 *mm* and 107 *mm* before the target position and having a spatial resolution of 1 *mm* (see fig.1). This assures the reconstruction of the beam track with a geometrical angular resolution of 0.14°. The emission angles of the tracks of the ejectiles were then reconstructed with respect to the measured beam track to improve the angular resolution (see fig. 1).

Some details of the two experiments can be found in the table 1.

The different energy resolution achieved in the two experiments can be observed in fig. 3 where the bare nuclear cross section (σ^N), free of Coulomb suppression effects, of the ^{18}F(p,α)^{15}O reaction is reported. The black dots report the data measured at CRIB, while the blue open squares are from the TAMU experiment. The full line is a multigaussian fit performed using data from the RIKEN experiment. The energies of the populated states of the ^{19}Ne are reported in the figure. The full line is a multigaussian fit performed on data from the RIKEN experiment. The energies in keV of the populated states of the ^{19}Ne are reported in the figure. The width of each gaussian was fixed by using the experimental resolution calculated by using a Monte Carlo simulation. A very interesting application of the THM to RIBs induced reactions is the study of neutron induced reactions on radioactive isotopes, even if they have a short lifetime [7–9]. Preliminary analysis of the ^{18}F(n,α)^{15}N has been

Table 1. Experimental details of the experiments performed at RIKEN- Japan and TAMU - USA for the measurement of the $^{18}F(p,\alpha)$ reaction

	CRIB - RIKEN - Japan	Mars - TAMU - USA
	^{18}F beam	
Beam energy	47.9 MeV (FWHM 1.9 MeV)	52 MeV (FWHM 2.6 MeV)
Beam spot	tracked	$3\ mm \times 5\ mm$
Beam purity	98%	94%
Beam intensity	5×10^5 pps	3-4×10^5 pps
	CD_2 target	
target thickness	150-200 μg/cm^2	400-800 μg/cm^2
	Detectors	
angular range	2°-11°	3°-12°
	11°-31°	15°-40°
angular resolution	0.2°	0.7°
(geometrical)	0.2°	1.1°
energy resolution	0.8%	0.8%

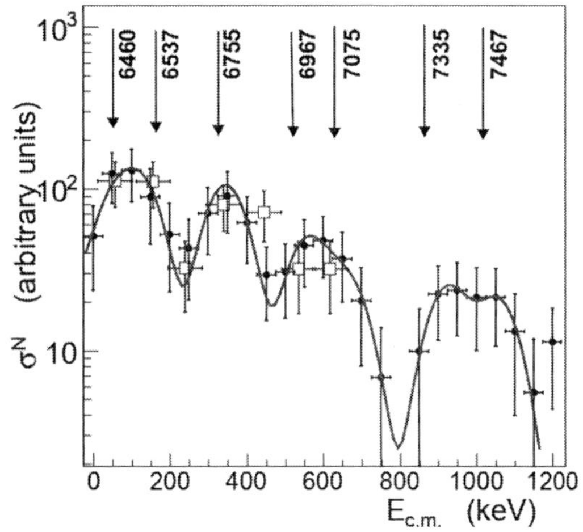

Figure 3. Bare nuclear cross section for the $^{18}F(p,\alpha)^{15}O$ reaction measured by using the THM. The full dots refer to the RIKEN experiment while the blue open square are from the TAMU experiment. The full line is a multigaussian fit performed on data from the RIKEN experiment. The energies in keV of the populated states of the ^{19}Ne are reported in the figure.

performed using the experimental data from the CRIB experiment. The extracted differential bare nuclear cross-section is shown in fig. 4.

3 Conclusions

The THM was extended to study reactions induced by RIBs. Even with the use of indirect methods like THM, the measurement of cross sections of interest for nuclear astrophysics remains one of the most difficult tasks in nuclear physics, especially if one has to use RIBs. Indeed, low beam intensity,

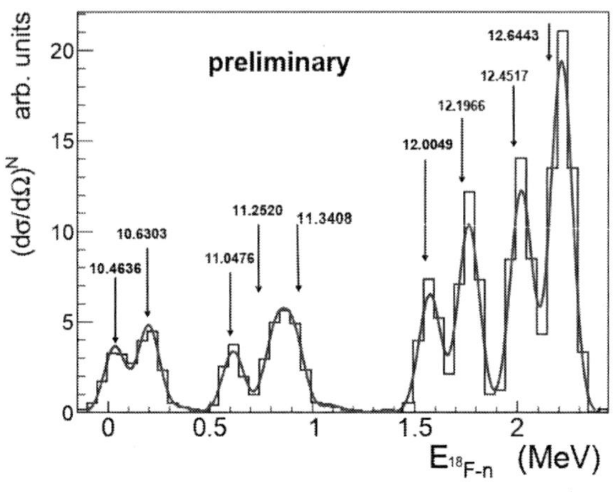

Figure 4. Bare nuclear cross section for the $^{18}F(n,\alpha)^{15}N$ reaction measured with the THM at the CRIB facility, RIKEN, Japan.

large divergence and presence of contaminants are added on the difficulties of measuring the low cross sections, which are typical of astrophysical nuclear processes.
The THM has been successfully used to infer pieces of information about the $^{18}F(p,\alpha)^{15}O$ by studying the three body reaction $^{18}F(d,\alpha\ ^{15}O)n$. Two experiments were performed in two different laboratories and using different experimental set-up, and similar results have been obtained. The comparison has stressed the fundamental role played by the beam tracking system to improve the energy resolution of the obtained cross section.
Calculations on the rate of the $^{18}F(p,\alpha)^{15}O$ reaction performed by using a modified R-matrix code have shown an higher destruction rate in the temperature range of interest for the novae. Consequently, the ^{18}F content in novae is decreased by a factor 2 [10].

References

[1] C. Spitaleri. in Proceedings of the Fifth Hadronic Physics Winter Seminar, Folgaria - TN, Italy, Ed. World Scientific, Singapore, 1990.
[2] S. Cherubini et al., ApJ **457**, 855, (1996)
[3] S. Cherubini et al., Phys. Rev. C **92**, 015805 (2015)
[4] R.G. Pizzone et al., EPJA, 52 24 (2016)
[5] G.G. Ohlsen NIM **37**, 240, (1965)
[6] B.T. Roeder et al., Nucl. Instrum. Methods A **634**, 71 (2011)
[7] M. Gulino et al., J. Phys. G: Nucl. Part. Phys. **37**, 125105 (2010)
[8] M. Gulino et al., Phys. Rev. C **87**, 012801 (2013)
[9] M. Gulino et al., Nuovo Cimento C - Coll. and Comm. in Phys. **39**(5), 369 (2016)
[10] M. La Cognata at al., ApJ **846**, 65, (2017)

Transfer reactions for nuclear astrophysics

Fairouz Hammache[1,⋆]

[1]*Institut de Physique Nucléaire d'Orsay, UMR8608, IN2P3-CNRS, Université Paris sud 11, 91406 Orsay, France*

Abstract. Direct measurements of cross sections at stellar energies are very challenging - if at all possible. This is essentially due to the very low cross-sections of the reactions of interest (especially when it involves charged particles), and/or to the radioactive nature of many key nuclei. Direct measurements with charged particles are often performed at higher energies and then extrapolated down to stellar energies using R-matrix calculations. However, these extrapolations are delicate because of the possible existence of unobserved low-energy or sub-threshold resonances. In order to bypass the difficulties related to direct measurements, indirect methods such as transfer reactions are used. These experiments are usually performed at higher energies and their conditions are relatively less stringent than in direct measurements. However, these methods rely on theoretical models for which the input parameters may be an important source of systematic uncertainties and thus need to be determined carefully. In this manuscript, a short overview on the difficulties related to direct measurements will be given as well as a description of the transfer reaction method and the theoretical concept behind. Finally, the method will be illustrated through two recent performed studies.

1 Introduction

Our understanding of stellar evolution in the Universe has been largely improved thanks to the interaction between three fields: observation, stellar modeling and nuclear physics. All these fields are in constant development: new telescopes and satellites open more and more windows on the cosmos, stellar modeling relies on ever-increasing computing and nuclear physics takes advantage of new facilities (radioactive beams high-intensity beams, underground laboratories) and sophisticated detection systems.

Nuclear reaction rates are one of the most important ingredients in describing how stars evolve. The study of the nuclear reactions involved in different astrophysical sites is thus mandatory to address most questions in nuclear astrophysics. Experimental techniques for cross sections determination lie in two main categories: direct measurements, in which the reaction of interest is reproduced, although the energy range may be different from the one in the stellar site; and indirect measurements, in which a different reaction is coupled with theoretical modeling to get the cross section of interest or to access the spectroscopic properties (Ex, J^{π}, decay widths,...) of the involved nuclei.

Direct measurements at stellar energies are very challenging - if at all possible. This is essentially due to the very low cross-sections of the reactions of interest (especially when it involves charged

⋆e-mail: hammache@ipno.in2p3.fr

particles), and/or to the radioactive nature of many key nuclei. Until recently, direct measurements with charged particles were often performed at higher energies and then extrapolated down to stellar energies using R-matrix calculations. However, these extrapolations are delicate because of the possible existence of unobserved low-energy or sub-threshold resonances. The effect of these resonances may change the extrapolated S-factor at astrophysical energies by a huge factor (sometimes orders of magnitude).

The other problem concerning the direct measurements is related to the radioactive nature of the nuclei involved in reactions occurring in explosive sites (novae, supernovae, X-ray bursts,...) and those involved in (n,γ) radiative captures in r-process and sometimes in s-process. The intensities of the radioactive beams are often low, rarely exceeding 10^5 to 10^6 pps and for nuclei with relatively long half life, making targets with enough atoms per cm^2 is very difficult. Hence the direct measurements of such reactions are very difficult and challenging and in case of r-process reactions it is currently impossible.

To bypass these difficulties (sub–threshold resonances, radioactive nuclei,...) indirect methods such as transfer reactions [1], Coulomb dissociation [2], ANC method [3] and Trojan Horse Method [4] are good alternatives. In these various methods, the experiments are usually performed at higher energies than the Coulomb barrier which imply higher cross sections than in direct measurements. Morevover these methods allow also to use stable beams to study reactions involving radioactive nuclei not far from the valley of stability. However, these methods are model dependent. They depend on the uncertainties relative to the different parameters used in the model. Hence, there are two sources of errors, experimental and theoretical. But the global uncertainty on the measured cross section can be reduced by combining different methods.

In this manuscript, I will focus on the the transfer reaction method. First of all I will show the possible type of reactions one can study using this method: the resonant one and the direct one. Then a description of the method and the theoretical concept behind will be given as well as some examples of experiments using this method.

2 Resonant and direct reactions

2.1 Resonant reactions

A resonant reaction, as shown in Fig. 1, is a two step process where the entrance channel A+x forms first an excited state E_x in the compound nucleus C at the incident energy E_{cm} and then the formed resonant state decays either to the exit channel B+y or to its ground state by emitting γ-rays. The resonant process can occur only at energies where the incident energy in center of mass is very close to the resonance energy which is given by: E_r=E_x-Q where Q is the Q-value of the reaction.

The cross-section of a resonance is given by the Breit-Wigner formulae:

$$\sigma(E) = \pi\lambda\!\!\!^{-2}\frac{2J_C+1}{(2J_A+1)(2J_x+1)}\frac{\Gamma_x\Gamma_y}{(E-E_r)^2+(\Gamma_{tot}/2)^2} \tag{1}$$

And in the case of a narrow resonance ($\Gamma_{tot} \ll E_r$), the reaction rate per particle pair is described by:

$$<\sigma v>=(\frac{2\pi}{\mu kT})^{3/2}(\omega\gamma)_R exp(-\frac{E_r}{kT}) \tag{2}$$

where the resonance strength is given by:

$$(\omega\gamma)_R = \frac{2J_C+1}{(2J_A+1)(2J_x+1)}\frac{\Gamma_x\Gamma_y}{\Gamma_{tot}} \tag{3}$$

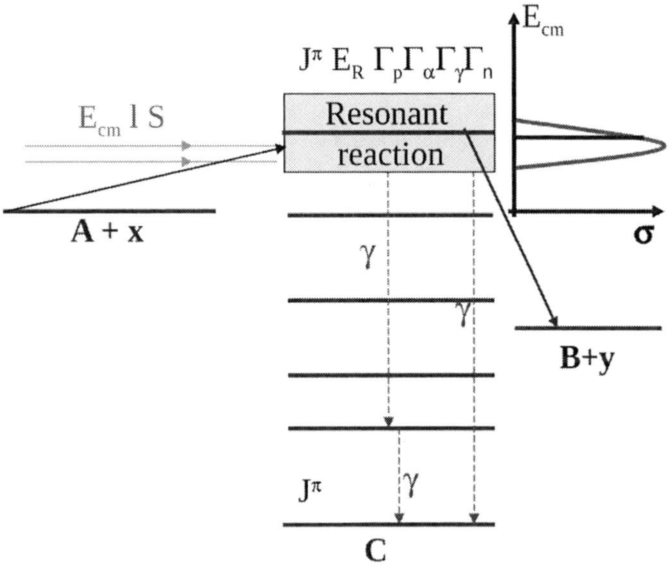

Figure 1. Resonant reaction scheme

The important parameters to describe a resonant cross-section and calculate the reaction rate are the resonance energy E_r, the partial width Γ_x which expresses the probability to form the compound nucleus C* in the excited state E_x, the partial width Γ_y which expresses the probability of decay to the exit channel B+y or C+γ and the spin-parity of the states involved in the compound nucleus J_C.

Once these different parameters are known, one can easily calculate the reaction rate using formulae (1). The excitation energy and the partial decay widths of the states of interest can be obtained from transfer reactions experiments.

Let's assume a compound nucleus C is in an excited state E_r (see Fig. 1) with a pure core-particle configuration $|A \otimes x\rangle$. Thanks to the time reverse invariance of the electro-magnetic and nuclear processes, the partial width of the formation of the compound nucleus in the excited state E_r through A+x is identical to the partial width of the decay of the excited state C* into A+x. The decay partial width of C into A+x is given by the following formulae:

$$\Gamma_x^{s.p} = \frac{\hbar^2}{\mu} R P_l(E,R) |\varphi(R)|^2 \tag{4}$$

with

$$\int_0^\infty \varphi(R)|^2 R^2 dR = 1 \tag{5}$$

Where R is is the channel radius, P_l is the penetrability factor, E is the energy of the free particle x and φ(R) is the radial wave function of the particle x. Hence, for a state with a pure core particle configuration, the single particle $\Gamma_x^{s.p}$ can be calculated. But in most cases, the states are not pure core-particle configurations, they are a mixture of configurations which decay or formation partial width Γ_x is given by the product of the single particle width Γ_x^{sp} and the spectroscopic factor S:

$$\Gamma_x = \Gamma_x^{sp} \times C^2 S \tag{6}$$

where C is the isospin Clebsch-Grodon coefficient.

The spectroscopic factor S expresses the overlap probability between the antisymmetrized wave functions of the entrance channel A+x and the final state C; S=$\langle A + x|C\rangle$. Hence, by determining the spectroscopic factor S via transfer reactions, one can then calculate Γ_x or the reduced width as we will see in section 3.

2.2 Direct (n,γ) captures

For reactions where direct capture mechanism can sometime plays an important role such as in ^{48}Ca(n,γ)^{49}Ca reaction [5], the capture occurs on bound states of the final nucleus in a one step process (see Fig. 2). The captures are possible at all neutron energies and the cross section varies smoothly with the neutron energy.

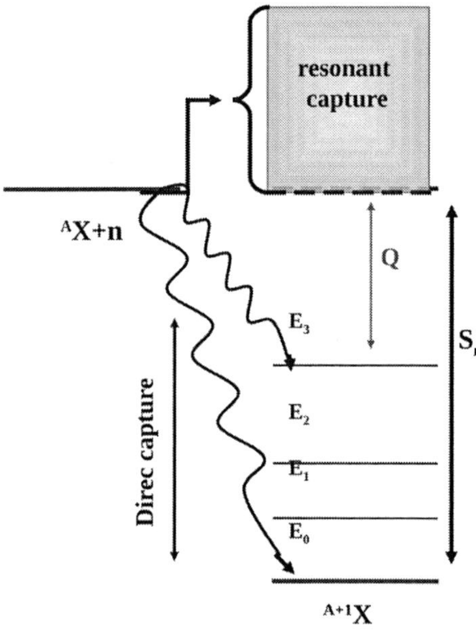

Figure 2. Direct capture reaction scheme

The cross section of the direct process is described by the following single matrix element [5]:

$$\sigma_{(n,\gamma)} = \sum_i C_i^2 S_i | \int_{r=0}^{\infty} \phi_f \theta_{em} \phi_i d\vec{r} |^2 \tag{7}$$

Where θ_{em} is the electro-magnetic multipole transition operator, ϕ_i and ϕ_f are the scattering and the bound state wave functions in the entrance and exit channels, respectively. The wave functions are Schrödinger's equation solutions with the potential V obtained by double folding and S is the spectroscopic factor of the final state. The double folding potential V is given by:

$$V = \lambda \int \int \rho_n(\vec{r}_n)\rho_A(\vec{r}_A)v_{eff}(E_n, \rho_n, \rho_A, |\vec{R} - \vec{r}|)d\vec{r}_n d\vec{r}_A \tag{8}$$

where v_{eff} is the nucleon-nucleon interaction , ρ_n is the neutron density, ρ_A is the nucleus target density and λ is a factor adjusted to reproduce the elastic scattering data for the entrance channel and the neutron separation energy for the exit channel. The latter depends on the location of the excited states in the final nucleus. The spectroscopic information on the low energy bound states (E_x, l, C^2S) are accessible via (d,p) transfer reactions in case of (n,γ) reaction.

3 Transfer Reaction Method

Let's assume that we want to study a resonant reaction x+A→C*→B+y and measure the partial decay width Γ_x of the state of interest in C* into the entrance channel A+x. To do so, we populate the excited states of C by transfering the small particle x (see Fig. 3) which can be a nucleon or a cluster of nucleons from the nucleus a to the nucleus A.

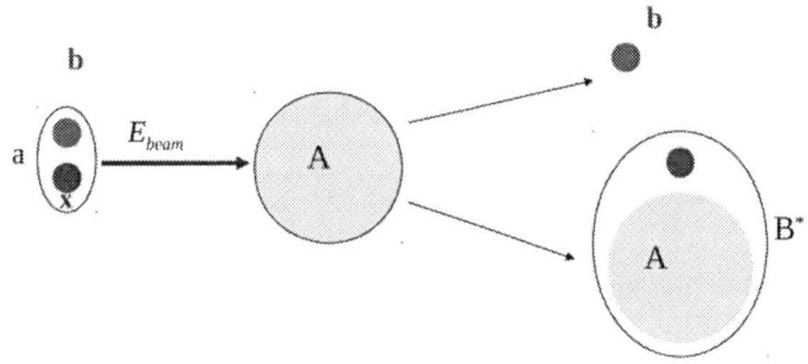

Figure 3. Sketch of a transfer reaction before (left) and after reaction (right).

This will feed the valence states of the final nucleus C, hopefully with no perturbation of the core, that is why it is called one step direct transfer reaction. The other part of the projectile b will continue its movement and will be detected. By measuring the emitted angle and energy of the particle b, one can deduce the energy of the excited state that was populated in C from kinematics and by comparing the shape of the measured angular distributions to those predicted by the distorted wave Born approximation theory (DWBA), one can extract the orbital angular momentum l of the populated single particle bound state. The j-values can also be obtained from polarization measurements.

The theoretical direct transfer cross section is calculated using the DWBA formalism and it is given by the following matrix element:

$$(\frac{d\sigma}{d\Omega})_{DWBA} \propto | < \chi_f I^C_{xA}(r_{xA})|V|I^X_{bx}(r_{bx})\chi_i > |^2 \tag{9}$$

Where $\chi_{i,f}$ are the distorted wave functions of the initial and the final state, V is the transition transfer operator, $I^{C}_{xA}(r_{xA})$ is the overlap function of the final bound state C formed by A+x and $I^{X}_{bx}(r_{bx})$ is the overlap function of the bound state a formed by b+x.

The radial part of these last functions is given by the following product:

$$I^{\alpha}_{\beta\gamma}(r_{\beta\gamma}) = S^{1/2}\varphi_{\beta\gamma}(r_{\beta\gamma}) \tag{10}$$

Where S is the spectroscopic factor, $\varphi_{\beta\gamma}(r)$ is the radial wave function of the bound state C or a with α being the final bound state C or the bound state a, β being the transfered particle x and γ being A or b respectively.

The spectroscopic factor of the different populated states in C can be extracted from the ratio of the measured differential cross section to the one calculated by DWBA:

$$(\frac{d\sigma}{d\Omega})_{exp} = S_1 S_2 (\frac{d\sigma}{d\Omega})_{DW} \tag{11}$$

As one can see in formulae 11, there are two spectroscopic factors, S_1 for the final bound state of interest in the exit channel and S_2 for the bound state in the entrance channel. Hence, the knowledge of the spectroscopic factor S_2 is mandatory to extract the the spectroscopic factor of interest S_1. Once the spectroscopic factor of the state of interest is extracted, one can then determine the reduced decay width using the following formula [6]:

$$\gamma_x^2 = \frac{\hbar^2 R}{2\mu} S_x |\varphi(R)|^2 \tag{12}$$

where $\varphi(R)$ is the radial wave function of the bound state C formed by A+x, calculated at a channel radius R where $\varphi(R)$ has its asymptotic behavior. The calculation of Γ_x is then given by:

$$\Gamma_x = 2P_l\gamma_x^2 \tag{13}$$

where P_l is the penetrability.

One can also extract the ANC [3] using the expression:

$$\tilde{C}^2 = S_\alpha \frac{R^2\varphi^2(R)}{\tilde{W}^2(R)} \tag{14}$$

where W(R) is the Whittaker function.

To calculate the transfer DWBA differential cross sections, a number of computer codes can be used such as FRESCO [7] and DWUCK [8]. In these codes, a number of inputs are needed and they have to be carefully determined. These inputs are the optical potential parameters describing the entrance channel, those describing the exit channel and finally the potential well parameters describing the interaction of the transfered particle with the core in the final and initial nucleus. For the optical potential parameters of the entrance and exit channel, elastic scattering measurements are needed. When elastic measurements are not available, one can use the potentials from measurements performed in the mass region close to the nuclei of interest and at close incident energies or the global potential formulae from Perey and Perey [9] abd Daehnick et al. [10]. Concerning the well parameters of the Woods-Saxon potential, the depth is adjusted to reproduce the binding energy of the bound state. For the radius and the diffuseness, different realistic sets can be used and the selected ones are those giving the optimal fit of the measured angular distributions. Note that the accuracy on the extraction of the spectroscopic factor can hardly be better than 20 %, so this method can not be used for reactions where an uncertainty better than 10% is needed.

The calculation of the transfer reaction cross sections with DWBA formalism involves the evaluation of a six dimensional integral [1]. For light-ion reactions such as (d,p) reaction, the direct reaction causing the transition is often approximated by a zero-range potential. This is partially justified by the small size of the light nucleus (e.g:deuteron) in comparison with the size of the other nuclei and the S wave state of the projectile. With this assumption, the DWBA integrals reduce to three-dimensional integrals that make the calculation simpler as only the form factor describing the interaction of the transfered particle with the core in the final nucleus has to be considered. However, this zero range assumption is no more valid if the projectile is not in an S wave internal state (L=0) and in case of reactions where the transferred particle is a cluster of more than 2 nucleons such as the α-particle. Here, the relatively large size of the nucleus (^{6}Li or ^{7}Li in case of α transfer) from which the cluster is transferred, together with the small wavelength associated with the relative motion, makes this assumption very poor and sometimes wrong. For these cases, finite range DWBA calculations have to be performed with a six dimensional integral which includes both form factors describing the interaction of the transferred particle with the core in the initial nucleus and in the final nucleus respectively.

When using the transfer method, one has to be aware that other reaction mechanisms such as the compound nucleus mechanism, the multi-step transfer reaction mechanism, the projectile breakup and the transfer to continuum [1] can occur. The contribution of these mechanisms can be evaluated by using Hauser Feschbach calculations, coupled channel calculations and continuum discretized coupled channel (CDCC) calculations respectively [7]. Note that usually the direct transfer mechanism is favored at small centre-of-mass angles ($\theta_{c.m}\leqslant 50°$), so one has to perform the transfer experiments at detection angles and incident energies where the compound nucleus component is reduced.

3.1 Examples of transfer reactions

Various type of transfer reactions can be used to study different reactions of astrophysical interest. The (^{3}He,d) and (d,n) proton transfer reactions are used to evaluate (p,γ) reactions by extracting the excitation energy, momentum and the proton spectroscopic factors of the states of interest as in Parikh et al.'s experiment [11]. The (p,d) and (p,t) transfer reactions are used to study proton-rich nuclei involved in reactions of astrophysical interest [12, 13]. For both of these reactions, the Q-values are strongly negative, hence proton beam with energies larger than 30 MeV are needed. Another interest of using (p,t) reactions comes from its selectivity, it populates only natural spin and parity states of even-even nuclei. To evaluate direct (n,γ) reaction cross-sections, one can use (d,p) neutron transfer reactions. In this case, one needs to use Adiabatic distorted wave approximation (ADWA) calculation which takes into account the deuteron breakup for deuterons energies larger than 20 MeV [14]. The α transfer reactions are very useful to study the properties of nuclei involved in α-induced reactions such as (α,γ) and (α,n) of astrophysical interest. The most commonly used are (^{6}Li,d) and (^{7}Li,t) reactions. From the Finite-range DWBA analysis of the measured angular distributions, one can extract the α spectroscopic factors and thus the ANCs and/or the α reduced widths of the states of interest needed to calculate the (α,n) and (α,γ) cross sections. According to previous works [6, 15, 16], the use of ^{7}Li instead of ^{6}Li reduces possible multi-step effects and the transfer cross sections to low spin states are enhanced because of the non-zero α angular momentum in ^{7}Li as observed when comparing the transfer reactions (^{6}Li,d) and (^{7}Li,t) on ^{12}C. These transfer reactions were applied to many astrophysical studies such as ^{12}C(α,γ)^{16}O [17], ^{13}C(α,n)^{16}O [18] and ^{22}Ne(α,n)^{25}Mg [19].

An illustration of the transfer method will be given in the following through two examples: first the study of the resonant reaction ^{13}C(α,n)^{16}O via a (^{7}Li,t) α-transfer reaction and the study of the direct component of ^{60}Fe(n,γ)^{61}Fe reaction via ^{60}Fe(d,p)^{61}Fe transfer reaction in inverse kinematics.

3.1.1 Application of the method to a resonant case: $^{13}C(\alpha,n)^{16}O$

Nearly half of the heavy elements observed in the Universe are produced by a slow sequence of neutron capture reactions, the so-called s-process. For this astrophysical process, two neutron sources are used in all evolution models: the reactions $^{13}C(\alpha,n)^{16}O$ and $^{22}Ne(\alpha,n)^{25}Mg$. For the AGB stars of 1-3 solar masses at low temperatures, only the$^{13}C(\alpha,n)^{16}O$ reaction is considered as the main neutron source for the s-process [20]. Hence, all the models describing the s-process nucleosynthesis in these AGB stars depend on the neutron flux from the $^{13}C(\alpha,n)^{16}O$ reaction and so on the cross section of this reaction which occurs in these stars at temperatures around 10^8 K, i.e around the Gamow peak E_{cm} ~190 keV.

Direct measurement of $^{13}C(\alpha,n)^{16}O$ reaction at this energy is extremely difficult because the cross section decreases drastically when the incident α energy decreases. Thus, direct measurements [21] have only been performed down to 260 keV too far away from the energy range of interest around 190 keV. R-matrix extrapolations [22] of the cross sections measured at higher energies have then to be performed, including the contribution of the 6.356 MeV, $1/2^+$, state of the compound nucleus ^{17}O, which lies 3 keV below the α+^{13}C threshold. The contribution of this sub–threshold state strongly depends on its α-spectroscopic factor, S_α. R-matrix extrapolations which consider a large value of 0.7 for the α spectroscopic factor (S_α) of the subthreshold state at 6.356 MeV lead to a rapid increase of the astrophysical S-factor when the energy decreases to zero while they lead to constant extrapolated S-factor at low energies when they use a null S_α.

Investigation of the effect of the sub–threshold resonance on the astrophysical S-factor was performed through a determination of the alpha spectroscopic factor of the 6.356 MeV state using the α-transfer reactions $^{12}C(^6Li,d)^{17}O$ [23, 24] and $^{13}C(^7Li,t)^{17}O$ [18]. The experiment [18] was performed using a $^7Li^{3+}$ beam of 34 and 28 MeV energy provided by the Orsay TANDEM impinging on a self-supporting enriched ^{13}C target. The reaction products were analyzed with an Enge Split-pole magnetic spectrometer (see Fig. 4) and the tritons were detected at various angles in the laboratory system.

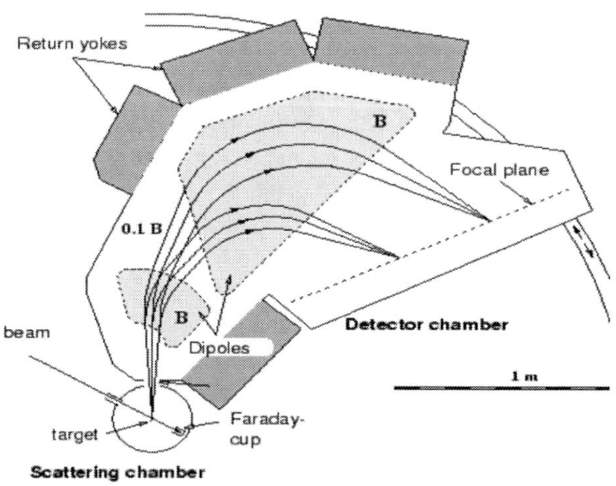

Figure 4. Schematic view of the Split-Pole spectrometer.

The measurements were performed at two incident energies in order to check the direct character of the transfer mechanism. Elastic scattering measurements were also performed at 28 MeV incident energy in order to get realistic optical potential parameters for the entrance channel. The measured elastic scattering angular distribution is displayed in Fig. 5.

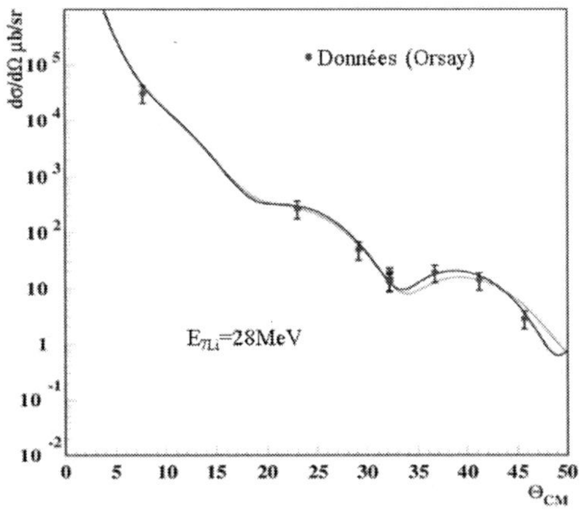

Figure 5. Experimental elastic $^{13}C(^{7}Li,^{7})^{13}C$ differential cross sections obtained at 28 MeV compared with DWBA calculations.

The experimental $^{13}C(^{7}Li,t)^{17}O$ differential cross sections measured for the 6.356, 3.055, 4.55 and 7.38 MeV, at the two incident energies of 34 and 28 MeV, are displayed in Fig. 6 together with the Finite-range DWBA calculations, using the FRESCO code [7]. See [18] for details.

Many combinations of entrance and exit optical potentials parameters were investigated. Concerning the entrance channel, several $^{7}Li+^{13}C$ optical potentials extracted from previous elastic measurements [25] at 34 MeV and the present elastic measurements at 28 MeV were tested. For the exit triton channel, optical potential parameters from Garrett et al. [26] were selected. The dependence of the calculation to the ^{13}C-α interaction potential was also investigated. A spreading of about 30% on S_α was found when the optical potential parameters were varied and another 20% with the variation of the well parameters. Finally, the selected optical and interaction parameters are those giving the best fit, using the maximum likelihood function set at 3σ level, for the studied states (6.05, 6.13, 6.92, 7.12 MeV) at both incident energies.

The α-spectroscopic factors were extracted from the normalization of the finite-range DWBA curves to the experimental data. The good agreement between the DWBA calculations and the measured differential cross sections of the different excited states of ^{17}O at the two bombarding energies of 28 MeV and 34 MeV respectively, gives strong evidence of the direct nature of the (^{7}Li,t) reaction populating these levels and confidence in the DWBA calculations. An S_α mean value of 0.29±0.11 is deduced for the sub–threshold state at 6.356 MeV in ^{17}O. The uncertainty on the extracted α spectroscopic factor for the state of interest 6.356 MeV was evaluated from the dispersion of the deduced S_α values at the two incident energies and using different sets of optical potentials in the entrance and exit channels and different α-^{13}C well geometry parameters [18].

The α-reduced width γ_α^2 of about 13.5±6.6 keV for the 6.356 MeV state was obtained using Eq.12. The calculation was performed at the radius R=7.5 fm where the Coulomb asymptotic behavior of the radial part of the α-^{13}C wave function is reached.

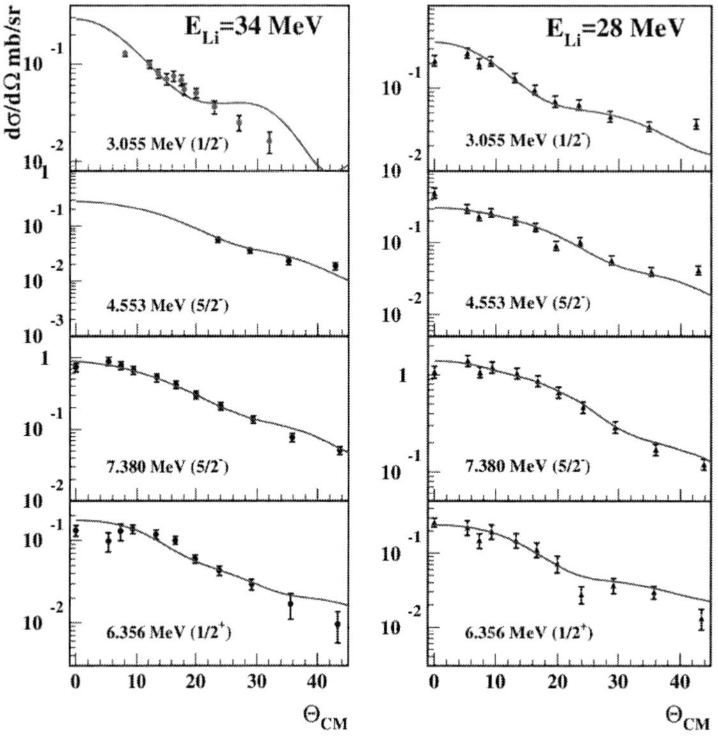

Figure 6. Experimental differential cross sections of the $^{13}C(^{7}Li,t)^{17}O$ reaction obtained at 28 and 34 MeV, compared with finite-range DWBA calculations normalized to the data.

The contribution of the $1/2^+$ state to the astrophysical S-factor when using the deduced γ_α^2 is shown in red curve in Fig. 7 At the energy of astrophysical interest, E_{cm}=0.19 MeV, the contribution of this sub–threshold state to the total S-factor was found dominant ($\approx$ 70%) [18]. This confirms the dominant character of the sub–threshold state on the cross section of $^{13}C(\alpha,n)^{16}O$ at the energies of astrophysical interest.

3.1.2 Application of the method to the non-resonant component of $^{60}Fe(n,\gamma)^{61}Fe$ reaction

The observation of ^{60}Fe characteristic gamma-ray lines by RHESSI and INTEGRAL spacecrafts [28, 29] as well as its observation in deep ocean crusts, sediments, lunar soils [30–32] and in galagtic cosmic rays (CRIS/ACE) [33] have underlined the need for accurate nuclear information concerning the stellar nucleosynthesis of this nuclei. ^{60}Fe is mainly produced in massive stars through the weak s-process component and it is released in the ISM by the sub-sequent core-collapse supernovae explosion [34]. Its production yield in core-collapse supernovae is uncertain due to the large uncertainties surrounding the cross sections of the production and destruction reactions $^{59}Fe(n,\gamma)^{60}Fe$ and $^{60}Fe(n,\gamma)^{61}Fe$ respectively.

$^{60}Fe(n,\gamma)^{61}Fe$ neutron capture reaction proceeds via two mechanisms, a resonant capture above the neutron threshold at 5.58 MeV and a direct capture on the bound states of ^{61}Fe below the threshold.

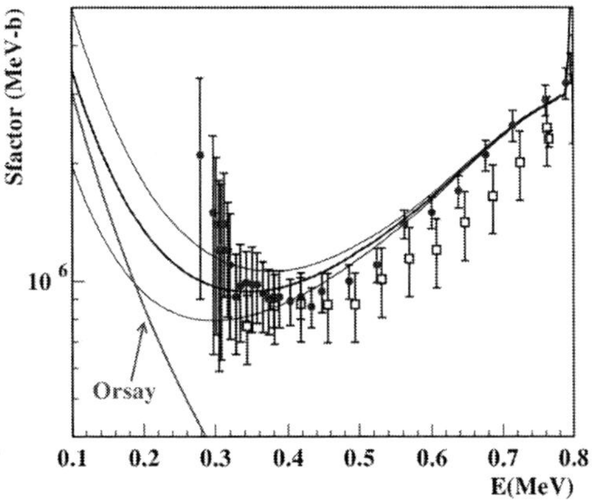

Figure 7. Astrophysical S-factor for the $^{13}C(\alpha,n)^{16}O$ reaction with R-matrix calculations. The data points are taken from Refs [21, 27]. The contribution of the 6.356 MeV state is shown as red curve. The thick black curve corresponds to the recommended γ_α^2 values, and the thin black ones to the lower and upper limits.

To evaluate the reaction rate, Hauser Feschbach calculations are used to estimate the resonant capture component while the direct component is determined using the equations described in section 2.2 and the spectroscopic informations (E_x, l, C^2S) of ^{61}Fe provided by shell model calculations since none of the measured states have known neutron spectroscopic factors.

As mentionned earlier to evaluate the direct component of an (n,γ) cross-section, one needs to determine the C^2S, Ex and l of the final bound states and (d,p) transfer reaction is the best tool to determine experimentally these parameters. An $^{60}Fe(d,p\gamma)^{61}Fe$ experiment was performed at GANIL in inverse kinematics [35] in order to check the reliability of the shell model calculations in the mass region N=34 and evaluate the contribution of the direct component to the total neutron capture cross section. An ^{60}Fe secondary beam was produced by fragmentation of ^{64}Ni on Be target at LISE spectrometer and then sent on a CD_2 target. Two MWPC called CATS were used to track the beam for proton emission point localisation. The energy and the emission angle of the emitted protons were detected at backward angles by four MUST2 telescopes and the annular detector S1 from micron. The emitted γ-rays from the decay of the populated states in ^{61}Fe were detected in 4 EXOGAM clovers. The aim of detecting the γ-rays is to disentangle the different populated states in ^{61}Fe that can not be discriminate with particle detection. Finally the quasi-projectiles were identified in mass and charge using the ToF and energy loss of the fragments in the ionsation chambre followed by a plastic scintillator.

The obtained ^{61}Fe excitation energy spectrum, using the MUST2 Silicon telescopes, is displayed in figure Fig. 8. Two peaks are observed below the neutron threshold and their width was found much larger than the expected energy resolution of about 800 keV. This is an indication of a population of two or more states in the observed peaks that the γ-rays measurements may help to disentangle.

From the comparison of the excitation energy spectrum with and without γ-ray coincidence and from the observation of the gamma-ray spectra corresponding to two energy gates in the first peak [35], 0 to 1 MeV and 1 to 2 MeV, three states were identified clearly: the known 207 keV, the 391 keV and the isomeric state at 861 keV. No other known states above 1 MeV [36] that were maybe

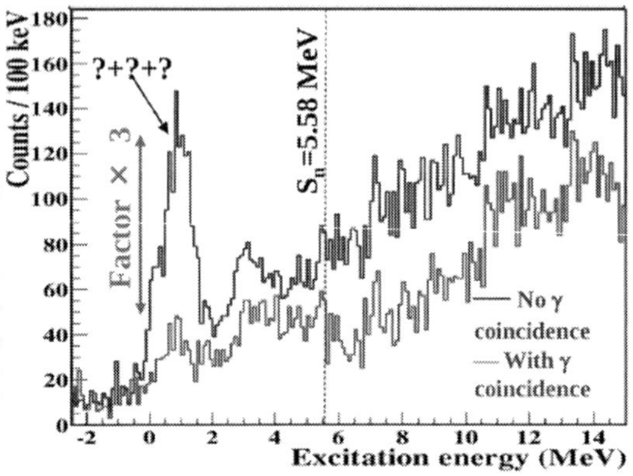

Figure 8. Measured excitation energy spectrum of ^{61}Fe with MUST2 telescopes

populated in this work could be identified reliably because of the large background and the low measured statistics due to the weak population of these states on one hand and the decrease of the γ-ray detection efficiency at high energies on the other hand.

A deconvolution of the first peak observed in ^{61}Fe excitation energy spectrum was performed with the ground state, the three well identified populated states in this experiment, the 207 keV (J^π=5/2$^-$), the 391 keV (J^π=1/2$^-$) and the 861 keV (J^π=9/2$^+$) and also a higher level centered at 1600 keV representing a mixture of the non-identified higher states between 1 MeV and 2 MeV. Proton angular distributions, displayed in Fig. 9 could be extracted for the 861 keV state and for the excitation energy range corresponding to 207 keV state where an overlap of the ground state and the 391 keV states is present. The red curves are Adiabatic distorded wave approximation (ADWA) calculations performed using Fresco code. A neutron spectroscopic factor of 0.34 was obtained for the f5/2 state from the fit of the data with a combination of an L=3 describing the 207 keV state and an L=1 describing the p1/2 of the 391 keV state and the p3/2 of the g.s for which the spectroscopic factors are fixed to the values 0.58 and 0.15 respectively obtained from the normalisation of the ADWA calculation to the single extracted data point at 4° from S1 telescope (see [35] for details). For the 861 keV state, a neutron spectrosopic factor of 0.38 was obtained from the normalisation of the calculations to the data.

A comparison between the neutron spectroscopic factors obtained in this work with those predicted in LNPS shell model calculations [37] shows a very good agreement between the experimental results and the calculations. This is an indication of the relaibility of these shell model calculations in the N=34 mass region.

The direct component of ^{60}Fe(n,γ)^{61}Fe was calculated using in TEDCA code [38] the extracted experimental neutron spectroscopic factors for the first four excited states of ^{61}Fe. A value of about 0.2 mbarn was obtained at 25 keV which represents only 2% of the total cross section.

4 Conclusions

In both resonant and non resonant capture reactions, the spectroscopic properties (J^π, E_r,C^2S (Γ)) of the involved states are needed to compute the reaction rates. We have seen that transfer reaction is a good alternative method that can be used to extract these useful spectroscopic informations. However, the optical potential parameters used for the DWBA calculations need to be carefully determined in

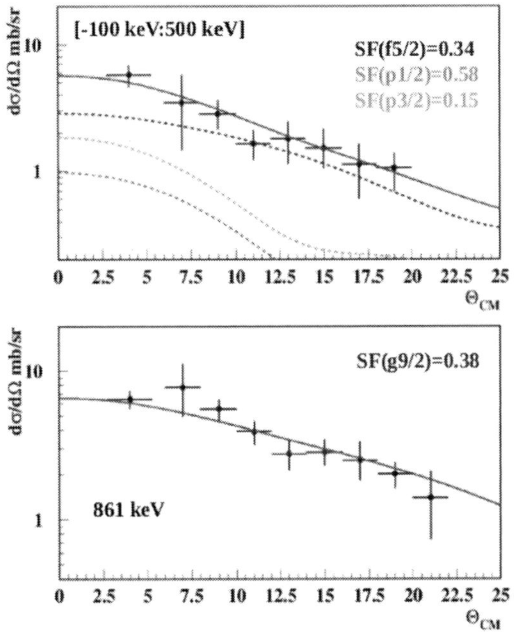

Figure 9. Experimental differential cross sections for a) the excitation energy range corresponding to the 207 keV state and b) the 861 keV excited state, together with the zero-range ADWA calculations normalized to the data. See text for details.

one hand and in the other hand one should also check the dominant character of the one step direct mechanism in the performed transfer reaction experiment.

References

[1] R. Satchler, *Direct nuclear reaction*, Clarendon Press, international series of monograohs on physics, 68 Ed., 1983
[2] G. Baur, C. A. Bertulani, H. Rebel, Nucl. Phys. **f 458**, 188 (1986).
[3] A. M. Mukhamedzhanov, R.E. Tribble, Phys.Rev.C **59**, 3418 (1999).
[4] C. Spitaleri et al., Phys.Rev.C **60**, 055802 (1999).
[5] E. Kraussmann et al., Phys. Rev. C **53**, 469 (1996)
[6] F. D. Becchetti and J. Jänecke, Nucl. Phys. A **305**, 293 (1978)
[7] I.J. Thomson et al., Comp. Phys. Rep. **7**, 167 (1988)
[8] P. D. Kunz spot.Colorado.edu/ kunz/DWBA.html
[9] Perey C. M. and Perey F. G. Atom. nucl. Data Tabl. **17** (1976)
[10] W. W. Daehnick et al., Physical Review C **21**, 6 (1980)
[11] A. Parikh et al., Phys. Lett. B **737**, 314 (2014)
[12] Z. Ma et al. Phys.Rev.C **76**, 015803 (2007).
[13] Bardayan et al. Phys.Rev.C **65**, 032801 (2002).
[14] G. L. Wales and R. C. Johnson, Nucl. Phys. **274**, 168 (1976)

[15] K. Kemper et al., Aust. J. Phys. **33**, 197 (1980)
[16] P. T. Debevec et al., Phys. Rev. C. **9**, 2451 (1974)
[17] N. Oulebsir et al., Phys. Rev. C **85** 035804 (2012) and references therein
[18] M.G. Pellegriti et al., Phys. Rev. C **77** 042801 (2008) and references therein
[19] R. Talwar et al., Phys.Rev. C 93, 055803 (2016) and references therein
[20] R. Gallino et al., Astrophys. J. **497**, 388 (1998)
[21] H. W. Drotleff et al., Astrophys. J. **414**, 735 (1993)
[22] P. Descouvemont, Phys. Rev. C. **36**, 2206 (1987)
[23] S. Kubono et al., Phys. Rev. lett. **90**, 062501 (2003)
[24] N. Keeley et al., Nucl. Phys. A **726**, 159 (2003)
[25] P. Schumacher et al., Nucl. Phys. A **212**, 573 (1973)
[26] J. D. Garrett et al., Nucl. Phys. A **212**, 600 (1973)
[27] C. R. Brune et al. Phys.Rev. C **48**, 3119 (1993)
[28] Smith et al., New Astr. ReV. **48**, 87 (2004)
[29] W. Wang et al., A and A **469**, 1005 (2007)
[30] K. Knie et al., Phys Rev Letter **93**, 171103 (2004)
[31] A. Wallner et al., Nature **532**, 69 (2016)
[32] L. Fimiani et al., Phys. Rev. Lett. **116**, 151104 (2016)
[33] W. R. Binns et al., Science **352**, 677 (2016)
[34] M. Limongi and A. Chiefi, New Astronomy Reviews **50**, 474 (2006)
[35] S. Giron et al., Phys.Rev. C **95**, 035806 (2017)
[36] D. Radulov et al., Phys.Rev. C **88**, 014307 (2013)
[37] S.M. Lenzi, F. Nowacki, A. Poves and K. Sieja, Phys. Rev. **82**, 054301 (2010)
[38] H. Krauss., Computer code TEDCA, (1995)

Experimental Approach to Explosive Hydrogen Burning in X-Ray Bursts and Core-Collapse Supernovae

Shigeru Kubono[1-3*]

[1]RIKEN Nishina Center, 2-1 Hirosawa, Wako, Saitama, 351-0198 Japan
[2]Institute of Modern Physics, Chinese Academy of Sciences, Lanzhou, China
[3]Center for Nuclear Study, the University of Tokyo, 2-1 Hirosawa, Wako, Saitama, 351-0198 Japan

Abstract. Recent experimental challenges to study the explosive hydrogen burning at extremely high temperatures are discussed. Along the nucleosynthetic pathway, discussions were made especially for the medium mass region, where we have possible waiting points and bottle necks which influence not only the light curve but also the production of heavy elements including the anomalously abundant p-nuclei at A = 80-100. Two approaches were discussed for the problem; one is the precision mass measurements and the other one is the half-live measurement of very short lived isotopes. A scope of the field is also discussed.

1 Explosive Hydrogen Burning at Extremely High Temperatures

Since hydrogen is the most abundant element, there are some sites where hydrogen burns explosively [1]. Typical such nucleosyntheses include the rapid proton-capture process (rp-process) in type I X-ray bursts, and the νp-process in type II supernovae. Type I X-ray burst will take place typically in a binary system involving a neutron star [2,3].

The property of neutron stars seems quite interesting, but is poorly known. Binary system of stars provide a good occasion to study the property of neutron stars. One is the type I X-ray burst, and other is the neutron star merger which was recently observed in August 2017, providing observations of the gravitational wave[4] and the following kilo nova phenomena [5]. The type I X-ray burst will take place by thermonuclear run away with accreted gas from the accompanying star onto the surface of the neutron star. It also provides an occasion to learn possibly the elemental evolution in the crust region of the neutron stars.

The second nucleosynthetic scenario, the νp-process, proposed in 2006 by three groups [6-8], is considered to take place in the early epoch of type II supernovae, where the site can be slightly proton-rich because of the mass difference of proton and neutron. This process is considered to produce very efficiently neutron-deficient nuclei in the light and medium mass regions. Thus, it is expected to make a significant contribution to the chemical evolution.

* Corresponding author: kubono@riken.jp

Therefore, this process is of great interest especially for production of the p-nuclei in the intermediate mass region (A = 80 ~ 100) [9], which are anomalously abundant isotopically. However, the nucleosynthetic flow path and the nuclide productions are largely uncertain because nuclear physics of the pathway region on the nuclear chart are still not known well yet [10,11]. Extensive mass measurements have been reported in the past decade, but not completed yet. Some examples are found in refs. [12,13].

Both the rp-process and the νp-process run through similar mass region in the neutron-deficient nuclear region, and have the common problems. However, the rp-process would have waiting points and bottle necks along the pathway, whereas the νp-process may have less effects at the points due to the neutron-induced reactions, (n,p) and (n,γ), which bypass the long-waiting beta-decays [10]. The waiting points also have a chance to be bypassed by the two-proton capture process (2p-process) [13] in both scenarios. Thus, the experimental challenges have been made in the past years to determine the rough pathway by precision mass measurements.

In the present lecture, I discuss recent experimental efforts around the possible waiting points in the rp-process, and production of the p-nuclei in the νp-process in the medium mass region.

2 Waiting Points at Intermediate Mass Region

There are many works reported for the nucleosynthesis at A < 56 in the rp-process, including experimental works using stable and radio-active nuclear beams [1]. Figure 1 illustrates the possible nucleosynthesis flow beyond ^{56}Ni, indicating possible waiting points and the bottle-necks. Sensitivity tests of each reaction along the rp-process pathway were also reported [14], suggesting crucial roles of nuclear reactions around the waiting points for heavy element synthesis, as shown in Fig. 1. The red squares in Fig. 1 are the possible waiting point nuclei, although none of them are well understood yet.

In this mass region, the proton capture reaction, A(p,γ)B, and the beta decay are the main processes for the rp-process. The reaction of A(p,γ) has a certain flow-back depending on the proton separation energy in nucleus B, i.e., whose relative contribution is defined by the Boltzmann factor, exp(-Q/*k*T), where Q is the proton separation energy, *k* is the Boltzmann constant, and T is the temperature of the site in kelvin [15]. At higher temperature with high density, the 2p-process may proceed to make a nucleus A+2p. For

Fig. 1. The rp-process flow predicted, where t red squires indicat possible waiting points.

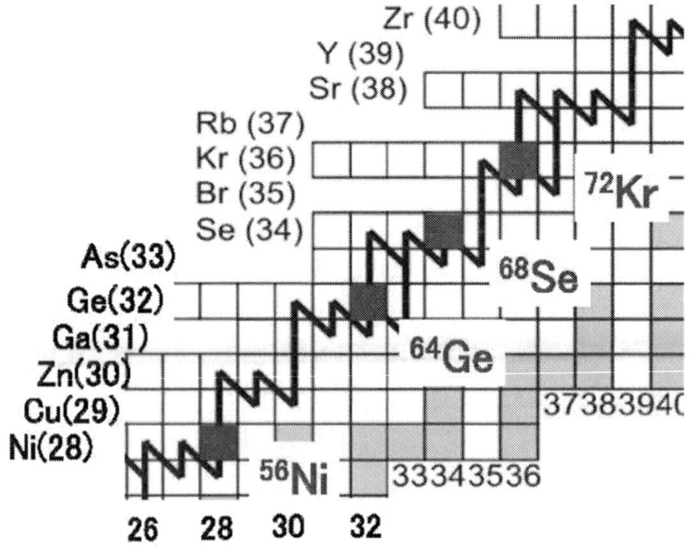

instance at ^{72}Kr in the figure, the beta decay of ^{72}Kr would compete with the 2p process. If the 2p process is very slow, then the rp-process flow has to wait for the beta decay to ^{72}Br. Since the half-life of ^{72}Kr is 17.3 sec. and the ordinary type I X-ray burst duration time is 20-30 sec, a large portion of the rp-process flow will stall at this point. To this problem, one needs to know the masses of the surrounding nuclei, and also needs to learn the proton capture process at the bottleneck points ^{73}Kr and ^{74}Rb.

2.1 Mass measurement of ^{65}As for the possible waiting point at ^{64}Ge

As discussed above, the masses of atomic nuclei play a crucial role for determining the basic path around the waiting points. High precision mass measurements have been made extensively for the nuclei along the possible pathway of the rp-process, although there still remain unmeasured nuclei. Such high precision mass measurement can be made using an ion storage ring, a Penning trap and an MR-TOF devices.

The first high precision mass measurement of ^{65}As was made at the CSRe ion storage ring at Lanzhou, China. The revolution time of each ions is described by the following equation [16];

$$\frac{\Delta T}{T} = \frac{1}{\gamma_t^2}\frac{\Delta(m/q)}{(m/q)} - \left(1 - \frac{\gamma^2}{\gamma_t^2}\right)\frac{\Delta v}{v}$$

The high precision mass measurement will be made by isochronous mass spectrometry, where the ions stored in a storage ring keep the isochronous condition by setting $\gamma = \gamma_t$ with γ_t the transition energy of the ring. Then, the precise m/q value will be extracted using reference particles whose masses are known with high precisions. Here, ^{65}As nuclides were produced by the fragmentation of ^{78}Kr at 480 AMeV, and injected into the CSRe ring together with other ions.

Figure 2 shows a time spectrum obtained in the measurement of ^{65}As, where the revolution time of ^{65}As was measured simultaneously with other ions. Experimental result obtained for the mass excess of ^{65}As is -46937 (85) keV. This gives the proton separation energy of -90 (85) keV, indicating the nucleus ^{65}As is slightly unbound for proton. Thus, the 2p process would be easier to go through from ^{64}Ge than thought before.

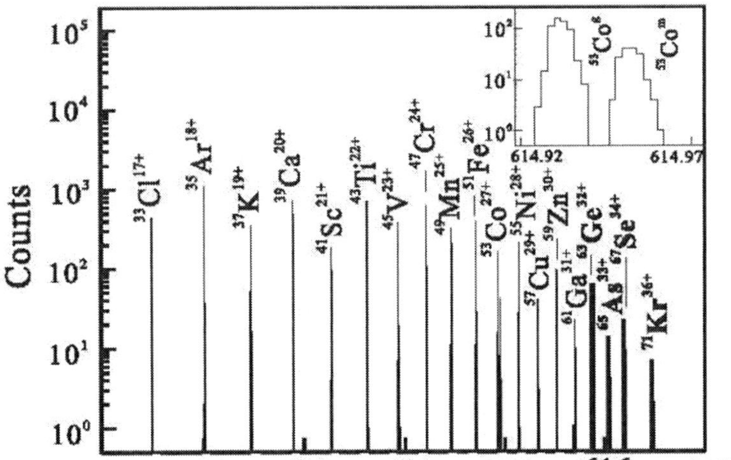

Fig. 2. The revolution time spectrum for the ^{65}As meassurement [12].

2.2 Possible waiting point at ^{72}Kr

Recently, an interesting result was obtained for the possible waiting point at ^{72}Kr in the EURICA campaign experiment at RIKEN-RIBF [17]. Figure 1 suggests that ^{72}Kr could be a waiting point nucleus for the rp-process. The rp-process flow would be greatly reduced if the 2p-process from ^{72}Kr is not favoured, as the half-life of ^{72}Kr is comparable to the decay time of the X-ray bursts, as discussed earlier. Figure 3 shows the yield distributions of N=35, 36 isotones including ^{73}Rb. A smooth distribution was

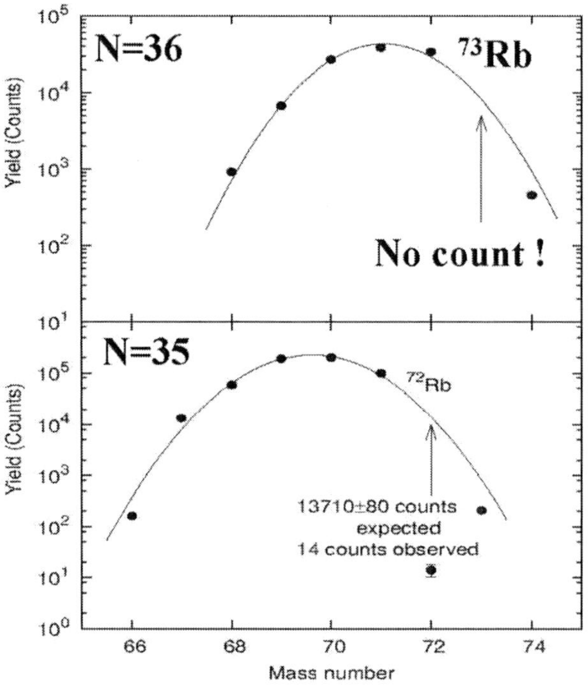

Fig. 3. The yields of isotones of N= 35 and 36. 73Rb was not observed at the experimental target station of EURICA [17].

observed except for ^{73}Rb in N=36. Since the measurements were made at the end of the zero-degree spectrometer of RIBF, about 130 m from the production target, ^{73}Rb is considered to have decayed to ^{72}Kr by a proton emission. The upper limit of the half-life was estimated to be 81 ns. The proton decay probability depends on the decay energy as well as the angular momentum of the shell orbit of the decaying proton in ^{73}Rb. Considering these facts together with possible deformation, one can constrain the proton decay energy. Here, the charged particle decay property is defined by the Coulomb penetrability as well as the centrifugal barrier penetrability for low energy proton decay. The estimated proton decay energy is not smaller than about 600 keV. Thus, the 2p-process seems difficult to proceed, suggesting the significant flux might stall at ^{72}Kr, although detailed model analysis is needed to conclude on this point.

3 p-Nucleus Production and the Termination Process

Haettner et al. [13] had reported precision mass measurements of the nuclei near the rp-process path at A=81-89, and claimed that the rp-process would terminate at around this mass region, because the ZrNb cycle would be formed, and thus the rp-process flow would

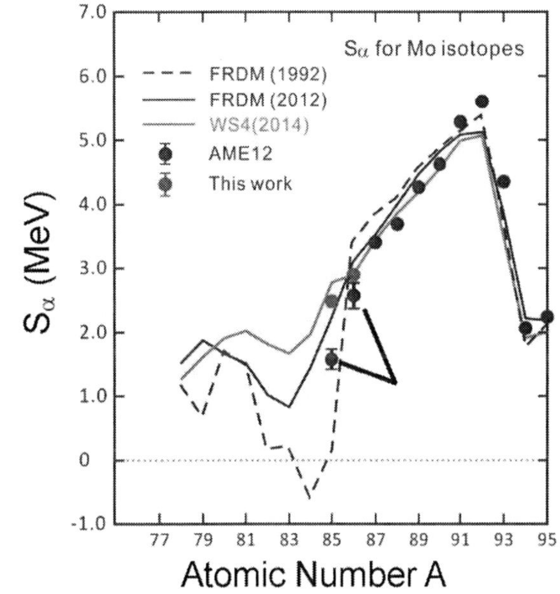

Fig. 4. The alpha separation energies were calculated based on the new masses determined at Lanzhou. The new data shows a smooth decrease suggesting the ZrRb cycle[13] would be less effective. [18].

terminate at this point at a certain temperature. However, this conclusion was found to be very uncertain because their analysis included some other masses not measured [13].

Recently, the masses of ^{82}Zr and ^{84}Nb were measured with high precision for the first time, and those of ^{79}Y, ^{81}Zr, and ^{83}Nb were re-measured at the CSRe storage ring in Lanzhou. Figure 4 shows the alpha separation energy derived based on the new masses, indicating the ZyNb cycle seems suspicious [18].

These new mass data of Lanzhou have a considerable impact to the p-nuclei production in this mass region, especially for production of ^{84}Sr. The νp-process was proposed by three groups [2 - 4], and all of them predict quite a large abundance of ^{84}Sr, but now with the new masses including ^{82}Zr, the simulation of the νp -process show significantly smaller abundance for ^{84}Sr. The new νp-process calculation produces less flow to ^{84}Mo, favouring a flow to ^{82}Y at ^{82}Zr, resulting in less ^{84}Sr production due to the new masses. Thus, the overproduction of ^{84}Sr was mainly originated due to lack of detailed mass information of the nuclei around A=84. This demonstrates that precision mass determination is very crucial for studying the p-nucleus production in the νp-process.

4 Summary and Scope

Explosive hydrogen burning process is one of the hottest subjects both for astronomical observation as well as for nuclear physics. Since nuclear reactions runs through proton-rich nuclear region in the explosive hydrogen burning process, this process has a great chance to contribute for p-nucleus production in the medium mass region. The most critical, physical parameters, the masses of nuclei along the possible pathway have been extensively measured, but still are not yet fully measured with precision. In the coming years, not only masses of nuclides, but also the proton capture processes at the bottle necks would become the next important subject for experiments to get full nuclear physics information to understand the hydrogen burning condition, and to know the contribution for p-nuclei.

As for the νp-process, one another important subject is to study the neutron induced reactions around the waiting point nuclei [11]. The (n,p) and (n,γ) reactions would discard the waiting points and accelerate the burning flow toward heavier mass region. These reactions are totally unknown.

The mechanism of type I X-ray burst will be further investigated not only by nuclear physics experiment, but also by astronomical observation which will tell us directly the heavy element synthesis in the bursts.

References

1. Proceedings of *the 14th International Symposium on Nuclei in the Cosmos NIC2016*, JPS Conference Proceedings No. **14** (2017) ed. S. Kubono, T. Kajino, S. Nishimura, T. Isobe, S. Nagataki, T. Shima, and Y. Takeda
2. R. K.Wallace and S. E.Woosley, Astrophys. J. Suppl. Ser. **45**, 389 (1981)
3. H. Schatz, et al., Phys. Rev. Lett. **86**, 3471 (2001)
4. B.P. Abbott, et al., Phys. Rev. Lett. **119**, 161101 (2017)
5. L.-X. Li and B. Paczn/sky, Astrophys. J. Lett. **509**, L59 (1998)
6. C. Fröhlich et al., Phys. Rev. Lett. **96**, 142502 (2006)
7. S. Wanajo, Astrophys. J. **647**, 1323 (2006)
8. J. Pruet et al., Astrophys. J. **244**, 1028 (2006)
9. M. Arnould and S. Goriely, Phys. Rep. **384** (2003) 1
10. S. Wanajo, H.-T. Janka, and S. Kubono, Astrophys. J. **729**, 46 (2011)
11. S. Kubono. Proc. *8th European Summer School on Experimental Nuclear Astrophysics*, IOP Publishing Journal of Physics: Conference Series **703,** 012009 (2016)
12. X.L. Tu et al, Phys. Rev. Lett. **106**, 112501 (2011)
13. E. Haettner et al., Phys. Rev. Lett. **106**, 122501 (2011)
14. R.H. Cyburt, et al., Atrophys. J. **830**, 55 (2016)
15. C.E. Rolfs and W.S. Rodney, *Cauldrons in the Cosmos*, The Univ. Chicago Press, 1988
16. X.L. Tu, et al., Nucl. Instr. Method A **654**, 213 (2011)
17. H. Suzuki, Phys. Rev. Lett. **119**, 192503 (2017)
18. Y.H. Zhang, et al., Proceedings of *the Origin of Matter and Evolution of Galaxies 2017*, Daejeon, Korea, 2017, to be published in AIP Conf. series

EPJ Web of Conferences **184**, 01011 (2018) https://doi.org/10.1051/epjconf/201818401011
9th European Summer School on Experimental Nuclear Astrophysics

Primordial Nucleosynthesis: Constraints on the Birth of the Universe

Grant Mathews[1,*], *Motohiko* Kusakabe[1,2], *Mayukh* Gangopadhyay[1,3], *Toshitaka* Kajino[2,4,5], and *Nishanth* Sasankan[1]

[1] *Center for Astrophysics, Department of Physics, University of Notre Dame, Notre Dame, IN 46556 USA*
[2] *School of Physics and Nuclear Energy Engineering, and International Research Center for Big-Bang Cosmology and Element Genesis, Beihang University, Beijing 100191, P.R. China*
[3] *Theory Division, Saha Institute of Nuclear Physics, 1/AF, Bidhannagar, Kolkata- 64, India*
[4] *National Astronomical Observatory of Japan 2-21-1 Osawa, Mitaka, Tokyo, 181-8588 Japan*
[5] *Department of Astronomy, Graduate School of Science, The University of Tokyo, 7-3-1 Hongo, Bunkyo-ku, Tokyo, 113-0033 Japan*

Abstract. We review the basic elements of big bang nucleosythesis (BBN) and how a comparison of predicted light-element abundances with observations constrains physics of the radiation-dominated epoch. We then summarize some applications of BBN and the cosmic microwave background (CMB) to constrain the first moments of the birth of the universe. In particular, we discuss how the existence of higher dimensions impacts the cosmic expansion through the projection of curvature from the higher dimension in the "dark radiation" term. We summarize current constraints from BBN and the CMB on this brane-world dark radiation term. At the same time, the existence of extra dimensions during the earlier inflation impacts the tensor to scalar ratio and the running spectral index as measured in the CMB. We summarize how the constraints on inflation shift when embedded in higher dimensions. Finally, one expects that the universe was born out of a complicated multiverse landscape near the Planck time. In these moments the energy scale of superstrings was obtainable during the early moments of chaotic inflation. We summarize the quest for cosmological evidence of the birth of space-time out of the string theory landscape. We will explore the possibility that a superstring excitations may have made itself known via a coupling to the field of inflation. This may have left an imprint of "dips" in the power spectrum of temperature fluctuations in the cosmic microwave background. The identification of this particle as a superstring is possible because there may be evidence for different oscillator states of the same superstring that appear on different scales on the sky. It will be shown that from this imprint one can deduce the mass, number of oscillations, and coupling constant for the superstring. Although the evidence is marginal, this may constitute the first observation of a superstring in Nature.

1 Introduction

We are at a unique period in the history of the human understanding of the cosmos. For the first time, we have a clear picture of what the universe is comprised of, how long it has been in existence, and

* e-mail: gmathews@nd.edu

how it will evolve in the future. This knowledge is the culmination of investigations via a number of cosmological probes including supernovae, observations of the large scale distribution of galaxies and the inter-galactic medium, analysis of the cosmic microwave background, observations of the first stars of the early universe, and studies of the nucleosynthesis of the elements in the first few moments of cosmic expansion. Here, we review a number of outstanding questions in contemporary cosmology and highlight the input that big bang nucleosynthesis (BBN) and the Cosmic Microwave Background provides toward answering them.

1.1 Elementary Dynamics of Space-time

If one adopts the notion that the universe is homogeneous and isotropic on a large enough scale, then space-time intervals are given by the Lemaitre-Friedmann-Robertson-Walker metric,

$$ds^2 = -dt^2 + a(t)^2\Big[\frac{dr^2}{1-kr^2} + r^2 d\theta^2 + r^2 \sin^2\theta d\phi^2\Big] \ , \qquad (1)$$

where now $a(t)$ is the dimensionless scale factor and k is the curvature parameter and we adopt units of $\hbar = c = 1$. For this metric, the evolution of the early universe is simply given by the Friedmann equation which describes the the Hubble parameter H in terms of densities ρ, curvature k, the cosmological constant Λ, and the cosmic scale factor a:

$$\left(\frac{\dot a}{a}\right)^2 \equiv H^2 = \frac{8}{3}\pi G\rho - \frac{k}{a^2} + \frac{\Lambda}{3} = H_0^2\Big[\frac{\Omega_\gamma}{a^4} + \frac{\Omega_m}{a^3} + \frac{\Omega_k}{a^2} + \Omega_\Lambda\Big] \ , \qquad (2)$$

where $H_0 = 67.74 \pm 0.46$ km sec^{-1} Mpc^{-1}is the present value [1] of the Hubble parameter. One can then define the various closure contributions from relativistic particles, nonrelativistic matter, curvature, and dark energy:

$$\Omega_\gamma \ = \ 8\pi G\rho_\gamma/(3H_0^2) \ , \Omega_m = 8\pi G\rho_m/(3H_0^2) \ , \ \Omega_k = k/(a^2H_0^2) \ , \ \Omega_\Lambda = \Lambda/(3H_0^2) \ . \qquad (3)$$

Based upon the *Planck* analysis [1] then one can deduce that the contribution to closure from baryonic matter is $\Omega_b = 0.0486 \pm 0.011$. The total matter content, however is $\Omega_m = 0.3089 \pm 0.0062$. This implies that a much larger fraction of the universe is made of a completely unknown component of "cold dark matter," $\Omega_c = 0.260 \pm 0.006$. Even more surprising is that the universe is predominantly made of a completely exotic form of mass energy, i.e. dark energy, denoted as, $\Omega_\Lambda = 0.691 \pm 0.006$ for the case of a cosmological constant. In addition to these, there is presently an almost negligible mass-energy contribution from relativistic photons and neutrinos which we designate as $\Omega_\gamma = \rho_\gamma/\rho_c = 5.46(19) \times 10^{-5.}$ The early universe includes the Planck epoch, the birth of space-time, inflation, reheating, a variety of cosmic phase transitions (e.g. supersymmetry breaking, baryogenesis, the electroweak transition, and the QCD transition), the epoch of big bang nucleosynthesis, and the production of the cosmic microwave background (CMB). The universe remains radiation dominated for most of the big bang, so that the radiation [Ω_γ term in Eq. (2)] is important during BBN. There are, however, interesting variants of big bang cosmology where this is not the case as described below.

The precision with which these parameters are now known is much better than a decade ago, however, one must keep in mind that t these parameters are based upon the simplest possible ΛCDM cosmology, and the analysis must be redone for more complicated cosmologies such as those described below.

The only direct probe of the radiation dominated epoch is the yield of light elements from BBN in the temperature regime from 10^8 to 10^{10} K and times of about 1 to 10^3 sec into the big bang. The only other probe is the spectrum of temperature fluctuations in the CMB which contains information of the first quantum fluctuations in the universe, and the details of the distribution and evolution of dark matter, baryonic matter, photons and electrons near the time of the surface of photon last scattering (about 2.8×10^5 yr into the big bang).

One of the most powerful aspects of BBN is the simplicity [2–7] of the equations. For the most part one can assume thermodynamic equilibrium of all species present. The only non-equilibrium processes are the thermonuclear reactions themselves. For all times relevant to standard big-bang nucleosynthesis, one can ignore curvature and dark energy. Hence, the Friedmann equation is just:

$$\left(\frac{\dot{a}}{a}\right) = \sqrt{\frac{8}{3}\pi G\rho_\gamma} = H_0\frac{\sqrt{\Omega_\gamma}}{a^2} \approx 1.13T_{\rm MeV}^2\ {\rm sec}^{-1} \ . \tag{4}$$

Also, at the time of BBN the timescale for Compton scattering is short. Hence, the number density of particles of type i with momenta between p and $p + dp$ is simply given by Fermi-Dirac or Bose-Einstein distributions,

$$n_i(p)dp = \frac{g_i}{2\pi^2}p^2\left[\exp\left(\frac{E_i(p)-\mu_i}{kT}\right) \pm 1\right]^{-1} dp \tag{5}$$

where $E_i(p)$ is the energy of the particle, the $\pm$ sign is negative for bosons and positive for fermions, while g_i is the number of degenerate spin states of the particle (e.g. $g = 1$ for neutral massless leptons, and $g_i = 2$ for charged leptons and photons).

Integrating the product $E_i(p)n_i(p)$ over all momenta gives the energy density for particle i. In the standard big bang, with all chemical potentials μ_i set to zero, the total mass-energy density of the universe at the epoch of nucleosynthesis is given by

$$\rho = \rho_\gamma + \rho_{\nu_i} + \rho_i \ , \tag{6}$$

where ρ_γ, ρ_{ν_i}, and ρ_i are the energy densities due to photons, neutrinos, and charged leptons, respectively (including antiparticles). Note that during BBN, the baryons contribute negligibly to the total mass energy density. At the temperatures of primordial nucleosynthesis only electrons (and positrons) contribute to the charged leptons. The contribution from neutrinos is given by (assuming only relativistic particles)

$$\rho_{\nu_i} = \int p[n_{\nu_i}(p) + n_{\bar{\nu}_i}(p)]dp \ = \frac{7}{8}\frac{\pi^2}{15}(T_{\nu_i})^4 \tag{7}$$

where T_{ν_i} is the neutrino temperature and here and for the rest of the manuscript we adopt natural units, i.e. $k = c = \hbar = 1$. The energy densities from photons ρ_γ and relativistic charged leptons ρ_i are given by

$$\rho_\gamma = \frac{4}{7}\rho_i = \frac{\pi^2}{15}(T_\gamma)^4 \ , \tag{8}$$

where T_γ is the photon temperature (henceforth we absorb the Boltzmann constant k into the temperature). For temperatures above 1 MeV, the neutrinos and photons are in thermal equilibrium ($T_\gamma = T_{\nu_i} = T$).

The Friedmann equation plus the fact that temperature decreases inversely with the scale factor, can be used to derive the relation between temperature and time during BBN.

$$T \approx 1.4g_{\rm eff}^{-1/4}\left(\frac{t}{1\ {\rm sec}}\right)^{-1/2}\ {\rm MeV} \ , \tag{9}$$

where $g_{\rm eff}(T)$ is the effective number of relativistic degrees of freedom in bosons and fermions,

$$g_{\rm eff}(T) = \sum_{\rm Bose} g_{\rm Bose} + \frac{7}{8}\sum_{\rm Fermi} g_{\rm Fermi} \ . \tag{10}$$

This completes the equations of the basic thermodynamics. The nuclear reactions, however, must be followed in detail as they fall out of equilibrium. For nuclide i undergoing reactions of the type $i + j \leftrightarrow l + k$ one can write:

$$\frac{dY_i}{dt} = \sum_{i,j,k} N_i \left(\frac{Y_l^{N_l} Y_k^{N_k}}{N_l! N_k!} n_k \langle \sigma_{lk} v \rangle - \frac{Y_i^{N_i} Y_j^{N_j}}{N_i! N_j!} n_j \langle \sigma_{ij} v \rangle \right) \tag{11}$$

where $Y_i = X_i/A_i$ is the abundance fraction, N_i is the number of reacting identical particles, n_i is the number density of nucleus i and $\langle \sigma_{ij} v \rangle$ denoted the maxwellian averaged reaction cross section,

$$\langle \sigma_{ij} v \rangle = \sqrt{\frac{8}{\pi \mu_{ij}}} (T)^{-\frac{3}{2}} \int_0^\infty \sigma_{ij}(E) \exp\left[-E/T\right] E \, dE \ , \tag{12}$$

where $\mu_{ij} = m_i m_j/(m_i + m_j)$ is the reduced mass of the reacting system. Once the forward reaction rate is known, the reverse reaction rate can be deduced from an application (cf. Ref. [8]) of the principle of detailed balance. For a two body reaction one can write:

$$\langle \sigma v \rangle_{lk} = \frac{g_i g_j}{g_l g_k (1 + \delta_{12})} \left(\frac{\mu_{ij}}{\mu_{kl}} \right)^{3/2} \langle \sigma_{12} v \rangle \exp\left[-Q/T\right] \ , \tag{13}$$

where Q is the energy released in the reaction. The reaction rates relevant to BBN have been conveniently tabularized in analytic form in several sources [9–11]. These rates are a crucial ingredient to BBN calculations.

In all there are only 16 reactions of significance during BBN. [5, 6, 12]. In order to be useful as a cosmological constraint one must know relevant nuclear reaction rates to very high precision ($\sim 1\%$). Fortunately, unlike in stars, the energies at which these reactions occur in the early universe are directly accessible in laboratory experiments. Although considerable progress has been made [6, 12–16] toward quantifying and reducing uncertainties in the relevant rates, improved reaction rates are still needed for the neutron life time [17, 18], the ^{2}H$(p,\gamma)^3$He, ^{2}H$(d,n)^3$He, ^{3}He$(d,p)^3$He, ^{3}He$(\alpha,\gamma)^7$Be, and ^{7}Be$(n,\alpha)^4$He reactions.

2 Light Element Abundances

One of the powers of standard-homogeneous BBN is that once the reaction rates are known, all of the light element abundances are determined in terms of the single parameter η (or η_{10} defined as the baryon-to-photon ratio in units of 10^{10}). The crucial test of the standard BBN is, therefore, whether the independently determined value of η_{10} from fits to the CMB reproduces all of the observed primordial abundances. There are new constraints on the primordial helium abundance [19]. The best abundance constraints for this review are then adopted from Ref. [6] as follows:

2.1 Deuterium

Deuterium is best measured in the spectra of narrow-line Lyman-α absorption systems in the foreground of high redshift QSOs. Moreover, i one restricts the data to the six well resolved systems for which there are multiple Lyman-α lines [6, 20], one obtains the presently adopted value [6] of $D/H = 2.53 \pm 0.04 \times 10^{-5}$.

2.2 ^{3}He

The abundance of ^{3}He is difficult to detect and the only measurement is from a single Galactic HII region [21] based upon the 8.665 GHz hyperfine transition of ^{3}He$^+$. A plateau with a relatively large dispersion with respect to metallicity was found at a level of ^{3}He/H = $(1.90 \pm 0.6) \times 10^{-5}$. It is not yet understood, however, whether ^{3}He has increased or decreased through the course of stellar and galactic evolution [22, 23]. Fortunately, one can avoid the ambiguity in galactic ^{3}He production by making use of the fact that the sum of (D +^{3}He)/H is largely unaffected by stellar processing. This leads to a reasonable 2σ upper limit [5, 13] of ^{3}He/H $< 1.7 \times 10^{-5}$.

2.3 ^{4}He

The primordial ^{4}He abundance, Yp is best determined from HII regions in metal poor irregular galaxies extrapolated to zero metallicity. In [19] it was demonstrated that updated emissivities and the neutral hydrogen corrections generally increase the inferred abundance, while the correlated uncertainties increase the uncertainty in the final extracted helium abundance. This leads to: $Y_p = 0.2449 \pm 0.0040$.

2.4 ^{7}Li

The primordial abundance of ^{7}Li is best determined from old metal-poor halo stars at temperatures corresponding to the Spite plateau (see Refs. [5, 6, 13] and Refs. therein). There is, however, an uncertainty in this determination due to the fact that the surface lithium in these stars may have experienced gradual depletion due to mixing with the higher temperature stellar interiors over the stellar lifetime. The best current limits are from Ref. [6]: ^{7}Li/H = $(1.6 \pm 0.3) \times 10^{-10}$.

3 Comparison of BBN with Observed abundances

Ultimately, the value of BBN is in the detailed comparison between the primordial abundances inferred from observation and the predictions based upon the BBN calculation. There is, however, one glaring problem that remains. The calculated and observed ^{7}Li/H ratio differ by about a factor of 3. This is known as "the lithium problem." A number of recent papers have addressed this problem [13, 14, 25–28]. At present it is not yet known if this discrepancy derives from a destruction of Lithium on the old stars used to deduce the primordial Lithium abundance, or if it requires exotic new physics in the early universe [13, 14, 25–27], or even a modification of the particle statistics in BBN itself [28]. For the remainder of this introduction we will adopt the premise that this disagreement may indicate new physics in the early universe. It is important then to address what are the key questions in modern cosmology and how might BBN address these questions.

3.1 Is there Evidence for Large Extra Dimensions?

In one form of the low-energy limit to M-theory [29] the universe can be represented by two 10-dimensional manifolds separated by a large extra dimension. It is possible that the extra dimension could manifest itself on the dynamics of the universe and BBN [30, 31]. For example, in a Randall-Sundrum II [32] brane-world cosmology, the cosmic expansion for a 3-space embedded in a higher dimensional space can be written [30, 31] as

$$\left(\frac{\dot{a}}{a}\right)^2 = \frac{8}{3}\pi G\rho - \frac{k}{a^2} + \frac{\Lambda}{3} + \frac{\kappa_5^4}{36}\rho^2 + \frac{\mu}{a^4} , \tag{14}$$

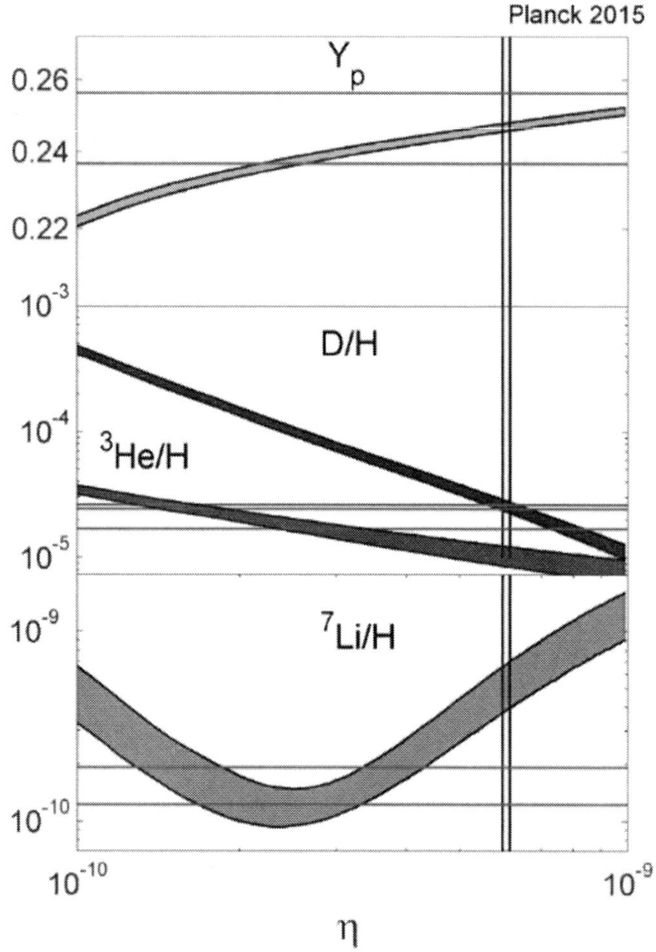

Figure 1. BBN abundances as a function of the baryon to photon ratio. Shaded bands correspond to the 2σ (95% C.L.) uncertainties deduced in [12]. Horizontal lines show range of the uncertainties in the primordial abundances inferred in [12]. Vertical lines indicate the value of η deduced in the *Planck* analysis [1].

where the four-dimensional gravitational constant G_N is related to κ_5 the five-dimensional gravitational constant, i.e. $G_N\kappa_5^4\lambda/48\pi$, with λ the intrinsic tension of the brane. The fourth term arises from the imposition of a junction condition for the scale factor on the surface of the brane, and is not likely to be significant during BBN. However it is significant during inflation [33] as described below. The fifth term, however, scales just like radiation with a constant μ and is called the *dark radiation*. Hence, it can contribute during BBN. Its magnitude and sign derives from the projection of curvature in higher dimensions onto four-dimensional space-time. Because this dark radiation scales as a^{-4} it can affect both BBN and the CMB. It can significantly alter [30, 31] the fit to BBN abundances and the CMB, and hence can be constrained. In [31] it was shown that the newest light element abundance constraints and reaction rates significantly limits the possibility of brane-world dark radiation.

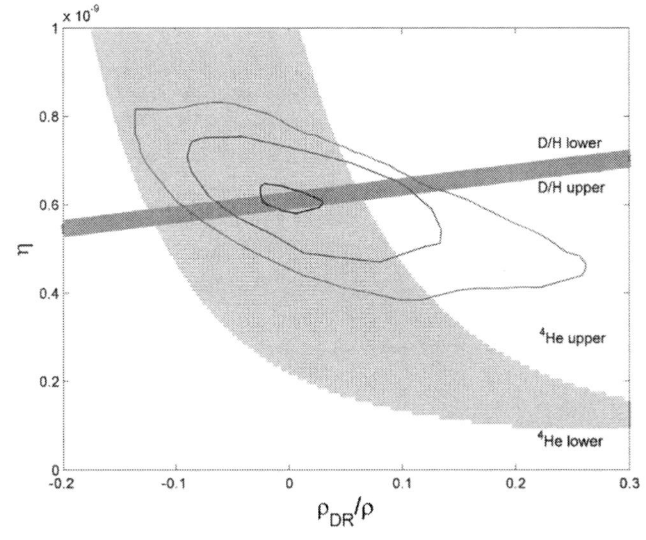

Figure 2. Constraints on dark radiation in the $\rho_{\rm DR}/\rho$ (10 MeV) vs. η plane from Ref. [31]. Contour lines show the 1, 2, and 3σ confidence limits based upon our fits to the CMB power spectrum. Dark shaded lines show the constraints from the primordial deuterium abundance as described in the text. The light shaded region shows the Y_p constraint.

Figure 2 shows the combined constraints on η vs. dark radiation based upon our fits to both BBN and the CMB power spectrum. Here, the fraction of dark radiation is normalized relative to the energy density ρ(10 MeV) at a big bang temperature of 10 MeV. The contour lines on Figure 2 show the CMB 1, 2, and 3σ confidence limits in the η vs. dark radiation plane. The shaded regions show the BBN Y_p and D/H constraints as labeled.

The best fit concordance shown in Figure 2 is consistent with no dark radiation although in the BBN analysis there is a slight preference for negative dark radiation. A similar result was found in the previous analysis of Ref. [30]. However, the magnitude of any dark radiation is much more constrained in the present analysis.

Similarly, in [33] we have analyzed the Randal Sundrum brane-world inflation scenario in the context of the latest CMB constraints from *Planck*. The simplest explanation for the fact that the universe is so nearly at today ($\Omega_0 = 1.000 \pm 0.005$ Ref. [1]) and the near isotropy of CMB is to conclude that the universe has gone through an epoch rapid inflation.[34] The simplest view is that some vacuum energy $V(\phi)$ drives inflation due to the existence of a self-interacting scalar field ϕ. That is, the energy density of the cosmic fluid in the early universe includes a dominant contribution from the evolution of a scalar field. The mass energy density for a scalar field can be deduced from the Klein-Gordon Equation $\rho(\phi) = \frac{\dot{\phi}^2}{2} + \nabla^2\phi + V(\phi)$. The inflaton field ϕ itself evolves according to a damped harmonic-oscillator-like equation of motion: $\ddot{\phi} + 3H\dot{\phi} - \nabla^2\phi + dV/d\phi = 0$.

As the universe expands, H is large and the $\dot{\phi}$ term dominates as a kind of friction term. The universe is then temporarily trapped in a slowly varying $V(\phi)$ dominated regime so that the scale factor grows exponentially. This is known as the slow roll approximation for which in the absence of spatial fluctuations becomes $\dot{\phi} = (dV/d\phi)/(3H)$.

The biggest unknown quantity in this paradigm is the form of $V(\phi)$. The simplest form $V(\phi) = (m/2)\phi^2$ may be motivated by the Kahler potential in string theory, however, almost any form for the potential works well to describe the big bang. Recently, the determination [1] ratio of the tensor to scalar contributions to the CMB power spectrum have ruled out many of the simplest forms of the inflation generating potential. In [33] we summarized constraints on the most popular classes of models and explore some more realistic inflaton effective potentials. The constraint on standard inflationary parameters changes in the brane-world scenario. We confirm that in general the brane-world scenario increases the tensor-to-scalar ratio, thus making this paradigm less consistent with the *Planck* constraints. However, in the case of the lowest order $\phi^{2/3}$ axion monodromy there can be a slight improvement over the standard model in the case of a large number of e-folds N, particularly when compared to the *Planck 15* constraints. Indeed, when BICEP2/Keck constraints are included, all monomial potentials in the brane-world scenario become disfavored compared to the standard scenario. However, for natural inflation the brane-world scenario fits the constraints better due to larger allowed values of e-foldings N before the end of inflation in the brane-world.

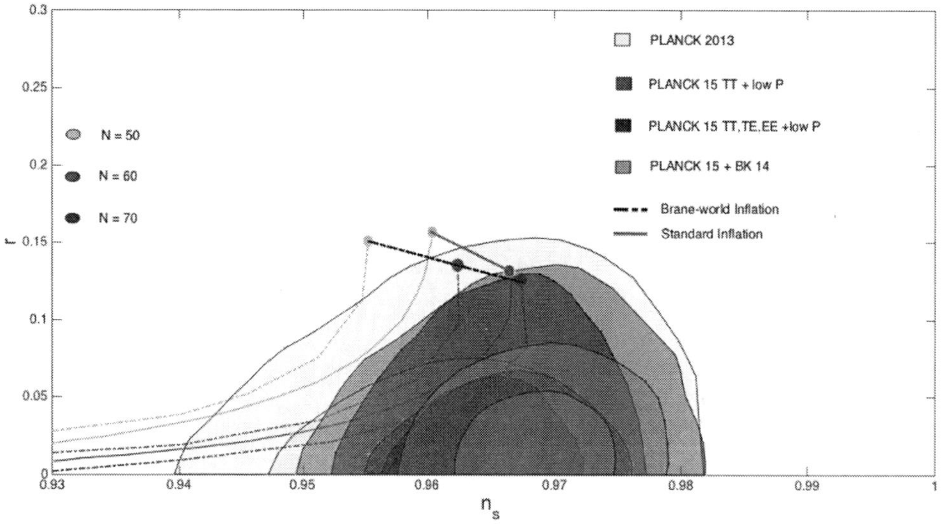

Figure 3. Effects of brane-world cosmology in the case of natural inflation compared to the standard inflation (from [33]). The legend remains the same as in previous figures. In this case, however, the dashed lines denote the range of solutions for different values of f in the brane-world scenario, while the solid lines show in the standard natural inflation solution.

For the most part, BBN is unaffected by the inflation-driving potential except in the special case of quintessential inflation [35]. In Ref. [35] we looked at this intriguing attempt to reduce the inflation potential problem, the baryogenesis question, and the dark energy mystery into a single paradigm which involves non-minimal coupling between matter and gravity as the universe makes a transition from an inflation driving potential to a dark-energy producing quintessence. Big bang nucleosynthesis significantly constrains this paradigm as the non-minimal couplings lead to an excess energy density in gravity waves which alter the results of BBN.

ACKNOWLEDGMENTS

Work at the University of Notre Dame is supported by the U.S. Department of Energy under Nuclear Theory Grant DE-FG02-95-ER40934.

References

[1] P. A. R. Ade et al., Planck XIII Collaboration, Astron. Astrophys. **594**, A13 (2016)
[2] R.V. Wagoner, Ap.J. **179**, 343 (1973)
[3] J. Yang, M.S. Turner, G. Steigman, D.N. Schramm, and K.A. Olive, Ap.J. **281**, 493 (1984)
[4] R. A. Malaney and G. J. Mathews, G. J., Phys. Rep. **229**, 145 (1993)
[5] F. Iocco et al., Phys. Rep., **47**, 1 (2009)
[6] R. H. Cyburt, B. D. Fields and K. A. Olive, T.-H. Yeh, Rev. Mod. Phys. **88** 015004 (2016)
[7] G. J. Mathews, M. Kusakabe and T. Kajino, Int. J. Mod. Phys. E **26**, 1741001 (2017)
[8] G. J. Mathews, Y. Pehlivan, T. Kajino, A. B. Balantekin and M. Kusakabe, Astrophys. J. **727** 10 (2011).
[9] R. H. Cyburt et al., Astrophys. J. Suppl. **189**, 240 (2010)
[10] G. R. Caughlan and W. A. Fowler, At. Data Nucl. Data Tables **40**, 283 (1988)
[11] C. Angulo et al., Nucl. Phys. A **656**, 656 (1999)
[12] M. Foley, N. Sasankan, M. Kusakabe and G. J. Mathews, Int. J. Mod. Phys. E **26**, 1741008 (2017)
[13] A. Coc and E. Vangioni, Int. J. Mod. Phys. E **26**, 1741002 (2017).
[14] R. Nakamura, M.-A. Hashimoto, R. Ichimasa and K. Arai, Int. J. Mod. Phys. E **26**, 1741003 (2017)
[15] P. Descouvemont et al., ADNDT, **88**, 203 (2004)
[16] C. Pitrou et al., Phys. Rep. *Submitted* (2018)
[17] A. P. Serebrov and A. K. Fomin, Phys. Rev. C **82**, 035501) (2010).
[18] G. J. Mathews T. Kajino, and T. Shima, Phys. Rev. D **71**, 021302 (2005)
[19] E. Aver, K. A. Olive, and E. D. Skillman, JCAP **05**, 003 (2010), JCAP **11**, 017 (2013); arXiv:1503:08146.
[20] R. Cooke et al., Astrophys. J. **781** 31 (2014); R. Cooke et al., Astrophys. J. **830**, 148 (2016)
[21] T. M. Bania, R. T. Rood and D. S. Balser, Nature **415**, 54 (2002)
[22] C. Chiappini, A. Renda and F. Matteucci, Astron. Astrophys. **395**, 789 (2002)
[23] E. Vangioni-Flam, K. A. Olive, B. D. Fields and M. Casse, Astrophys. J. **585**, 611 (2003)
[24] M. Kusakabe, K. Kim, M.-K. Cheoun, T. Kajino, Y. Kino and G. J. Mathews, Astrophys. J. Suppl. 214, 5 (2014)
[25] M. Kusakabe, G. J. Mathews, T. Kajino and M.-K. Cheoun, Int. J. Mod. Phys. E **26**, 1741004 (2017)
[26] J. Sato, T. Shimomura and M. Yamanaka, Int. J. Mod. Phys. E **26**, 1741005 (2017)
[27] D. Yamazaki, M. Kusakabe, T. Kajino, G. J. Mathews and M.-K. Cheoun, Int. J. Mod. Phys. E 26, 1741006 (2017)
[28] S. Q. Hou, J. J. He, J. J. A. Parikh, D. Kahl, D. C. A. Bertulani, T. Kajino, G. J. Mathews and G. Zhao, Astrophys. J. **834**, 165 (2017)
[29] P. Horava and E. Witten, Nucl. Phys. B **460**, 506524 (1996); P. Horava and E. Witten, Nucl. Phys. B **475**, 94114 (1996)
[30] K. Ichiki, M. Yahiro, T. Kajino, M. Orito and G. J. Mathews, Phys. Rev. D **66**, 043521 (2002)

[31] N. Sasankan, M. Gangopadhyay, M. Kusakabe and G. J. Mathews, Phys. Rev. D **95** 083516 (2017); Int. J. Mod. Phys. E. **26**, 1740076 (2017)
[32] L. Randall and R. Sundrum, Phys. Rev. Lett. **83**, 3370(1999); Phys. Rev. Lett. **83**, 4690 (1999)
[33] M. R. Gangopadhyay and G. J. Mathews, JCAP, in press, arXiv: 1611.50123 hep/ph (2018)
[34] A. Starobinsky, Sov. Phys. J. Exp. Theor. Phys. Lett. **30**, 682 (1979); A. Guth, Phys. Rev. D **23**, 347 (1981); A. Linde, Phys. Lett. B **108**, 393 (1982);
[35] M. Yahiro, G. J. Mathews, K. Ichiki, T. Kajino and M. Orito, Phys. Rev. D **65**, 063502 (2002)

RIB production and related experiments at EXOTIC

Marco Mazzocco[1,2,⋆]

[1] *Dipartimento di Fisica e Astronomia, Universita di Padova, via F. Marzolo 8, I-35131 Padova, Italy*
[2] *INFN - Sezione di Padova, via F. Marzolo 8, I-35131 Padova, Italy*

Abstract. Radioactive nuclei have a very deep relevance in many astrophysical scenarios, from the Big Bang nucleo-synthesis to supernova explosions. Several Nuclear Physics laboratories around the world have been constructing large-scale facilities for the production of Radioactive Ion Beams (RIBs). The main production techniques, i.e. In-Flight and Isotope Separation On Line, which will be reviewed in this contribution. In particular, we will concentrate on the production of light weakly-bound RIBs at the facility EXOTIC, located at INFN-LNL (Italy) and we will describe the most recent experiments.

1 Introduction

What is the most beautiful painting you have ever seen in your life? If this question is asked to a Nuclear Physicist, the answer might be a little surprising for a standard pedestrian. In fact, there is a rather high probability that the answer could be "The Chart of The Nuclides", i.e. the pictorial representation of the isotopes which was (originally) introduced in the mid-Thirties by the Italian Giorgio Fea [1]. In this chart, each known isotopes occupies a squared box and all boxes are properly ordered according to the atomic number (in the vertical axis) and to the number of neutrons (in the horizontal axis). Each box contains basic information on the nuclear properties of the relative isotope, such as, for instance, natural abundance (if stable) or mean half-live (if radioactive), presence of metastable states, cross section for the bombardment of thermal neutrons and so on. The chart is generally color-coded according to the main (radioactive) decay mode. Black is reserved for stable isotopes, blue for β^- decays, red for β^+ decays and electron captures, yellow for α radioactivity, green for spontaneous fission, orange for proton decay and light blue for neutron emission.

A quick look at the nuclide chart immediately gives rise to a first crucial question: what are the limits of existence of nuclei? The search for the heaviest particle-stable configuration of nucleons triggered the long-standing quest for the super-heavies [2]. On the other side, it is also extremely interesting to probe the limits of the nuclear stability along isobaric chains, i.e. let us fix the mass number and let us try to change one-by-one the number of protons and neutrons until we reach on both sides the last combinations stable against particle emission. These two extreme regions of the nuclide chart are called drip-lines and we have the proton- and neutron-drip line in the proton-rich and proton-deficient side the chart, respectively. Among these two lines and within the region of super-heavy nuclei, there is a "jungle" of 3532 isotopes [3] observed until the end of 2017, which is a sort of (radioactive) paradise for nuclear physicists and nuclear astrophysicists. However, our present-day

⋆e-mail: marco.mazzocco@pd.infn.it

knowledge of radioactive nuclei is still limited fo about 50-60% of the nuclei of possible existence. In fact, while the proton drip line has been essentially mapped up to Gold (Z = 79), the neutron-rich side is mostly an uncharted territory and this is region where, for instance, the path of the astrophysical rapid neutron capture process (*r*-process) is mostly located [4].

The interest in the properties of nuclei far from the valley of stability boosted the design and construction of large-scale Radioactive Ion Beams (RIB) facilities, such as for instance the RIBF at RIKEN (Japan) [5], ISOLDE at CERN (Switzerland) [6], ISAC at TRIUMF (Canada) [7], FRS at GSI (Germany) [8] and SPIRAL at GANIL (France) [9]. Besides these sites, there are also several small-scale facilities around the world and a few major RIB facilities under construction, such as FAIR at Darmstadt (Germany) [10], SPIRAL2 at GANIL (France) [11], SPES at INFN-LNL (Italy) [12] and FRIB at MSU (USA) [3].

In this contribution we will at first present the mostly used RIB production mechanisms, then we will discuss the two main RIB production techniques, In-Flight and Isotope Separation On Line (ISOL). Afterwards, we will describe the facility EXOTIC located at INFN-LNL (Italy) and, finally, summarize some of the most recent results obtained by means of this facility.

2 RIB production mechanisms

When a facility for the production of Radioactive Ion Bems is laid out, the following guidelines should be taken into account:

- the global production scheme should be fast, as the produced nuclear species are, in most cases, very short-lived and the number of radio-nuclides of interest decaying prior being investigated should be as small as possible;
- the production method should ensure the largest possible production cross section and this can be achieved by optimizing the projectile-target combination, the bombarding energy, the primary beam intensity, the power dissipation into the target, and so on;
- the production should efficient, as we are quite often dealing with very marginal production rates;
- the production should also be quite selective, since the contaminant species could be even several orders of magnitude more abundant than the nuclear species of interest.

There exist four main nuclear reaction mechanisms which can be employed for the production of RIBs, namely: fragmentation, fission, fusion and direct processes.

Fragmentation reactions are peripheral interactions leading to the removal of a few nucleons close to the nuclear surface. To be efficient the colliding energy should be larger than 40-50 MeV/u. In case the fragmentation is induced by an inverse kinematics reaction, i.e. an heavier projectile impinging on a lighter target, the residual nuclei are emerging from the interaction region with "small" recoils and are focused in narrow cones around the primary beam direction. Increasing the beam energy, i.e. building a more powerful driver accelerator, does not help enhancing the fragmentation cross reaction, it rather narrow the opening of the emission cones for the residues, thus diminishing the cost of the spectrometer to be used for the isotope selection. This technique is not extremely selective, can be employed to produce either proton-rich or neutron-rich side of the valley of stability and the projectile to be fragmented can, to some degrees of freedom, be tuned according to region of radioactive nuclei we are interested in.

The production of RIBs by means of the nuclear **fission** starts, of course, from the use either as projectile or target of a fissile material, typically ^{238}U. This process is mostly suitable for the production of neutron-rich nuclei, as it is quite understandable considering the extreme neutron-to-proton ratio of radio-isotopes unstable versus spontaneous fission. Fission fragments are emitted with energies (in the center-of-mass reference frame) of about 1 MeV/u and, in case of fissile projectiles, their kinematic cones around the primary beam direction have openings significantly larger than radioactive nuclei produced via fragmentation. In this case, building a more powerful accelerator can help achieving a much better overall production rate for very exotic fission fragments.

Nuclear **fusion** could be employed for the production of proton-rich nuclei. In this case, in fact, the compound nucleus is heavier than both projectile and target and, consequently, it will be characterized by a proton-to-neutron ratio much larger than stable isotopes of the same element. Recoil separators exploiting the substantial velocity difference between projectile and evaporation residues can be used to select the radioactive species produced. However, the separation efficiency is often quite limited, since the fusion process has the highest cross section at energies up to a few MeV/u and the residues are produced with broad angular, velocity and charge state distributions. So far, nuclear fusion has been used for spectroscopy studies, for the synthesis of super-heavy nuclei, but not for the production of RIBs.

Direct processes are generally employed for the production of light RIBs quite close to the valley of stability. Inverse (p,n), (d,n), or (^{3}He,n) reactions, preferably with negative Q-value, are chosen because of the narrow opening angles ($\theta_{lab} \leq 10°$) of the reaction products in the laboratory frame. Hydrogen and helium would be the preferred target materials. Unfortunately, there are no He-compounds and H-rich compounds (e.g., CH_2) cannot withstand the thermal stress induced by an intense primary beam intensity. To circumvent these problems, the use of gas targets with metal windows was undertaken.

3 RIB production techniques: In Flight vs. ISOL

In Flight separation and ISOL are considered complementary techniques for the production of RIBs.

In the **In Flight** production, an energetic heavy ion beam impinges on a thin light target. The target thickness is a crucial parameter, as the reaction products should retain most the projectile initial velocity (typically $\sim$ 90%) and the deflection introduced by the interaction with the target atoms should be as small as possible. The ions of interest are then selected by a proper combination of electromagnetic fields and atomic interaction. All the procedure is rather fast, as the radioactive ions are already at disposal of the experimentalists right after the transit time through the separator, at most after a few μs. Therefore, this technique is particularly suitable for short-lived RIBs. Moreover, since the RIB under production is very energetic, there is no need to equip the facility with a quite expensive post-accelerator. The drawbacks of this method are mostly related to poor optical properties of the outcoming RIBs, in terms of longitudinal and transverse emittance and beam spot on target, and to the limited flexibility on the secondary beam energy, strongly dependent on the energy of the primary beam.

On the other side, the **ISOL** technique is generally characterized by a light projectile beam, such as for instance a proton or a deuteron beam, impinging on a heavy target. The target station has a certain thickness, in order to stop completely the primary beam, and has to be kept at high temperature. In fact, the radioactive nuclei produced inside the target are thermalized in a catcher, then they are extracted, ionized, mass selected and eventually re-accelerated. Of course, all these different operations, especially those occurring inside the target/catcher container, critically depend on the chemical properties of the species under production and the extraction procedure, for instance, can require a time interval ranging from a few hundreds ms to some tens of seconds. Clearly, very short-lived nuclei are hardly accessible by means of the ISOL technique. However, the possibility of re-accelerating the secondary beam gives the opportunity of having an easy variable beam energy and also a high optical quality, i.e. very small beam emittance.

In the worldwide scenario, RIBF [5], FRS [8], FAIR [10]and FRIB [3] are In-Flight facilities, whereas ISOLDE [6], ISAC [7], SPIRAL (France) [9], SPIRAL2 [11] and SPES [12] are using (or will use) the ISOL technique.

4 The Facility EXOTIC at INFN-LNL

Our project for the design of a facility for the in-flight production of light weakly-bound RIBs at INFN-LNL started in 2002 [14]. We employ heavy ion beams, delivered by the 15 MV XTU-Tandem accelerator, impinging on a gas target which can be filled with light gases, such as ^{1,2}H and ^{3,4}He. The commissioning of facility was performed in 2004 [15–17] and a quite substantial upgrade process was held in 2012 [18]. The ion-optical lay-out of the beam-line consists of 3 quadrupole magnets, 1 30°-dipole magnet, a 1-m long Wien filter and finally a second quadrupole triplet. So far, seven RIBs have been delivered: ^{17}F, ^{8}B, ^{7}Be, ^{15}O, ^{8}Li and ^{10,11}C. In most cases, intensities of to a few units 10^5 pps and purities as good as 99% were achieved, except for ^{10}C (about 5×10^3 pps) and ^{8}B, where we never exceeded an purity of about 40% and we hardly reached a intensity of 10^3 pps. The facility is equipped with an event-by-event tracking system based on two x-y sensitive high-transparency Parallel Plate Avalanche Counters (PPACs) and an array of 8 $\Delta E - E$ telescopes of large-area Double Sided Silicon Strip Detectors (DSSSDs) is located [19, 20] at the final focal plane for energy and position measurement and for particle identification purposes.

These RIBs were used for reaction dynamics studies at Coulomb barrier energies for the systems ^{17}F + ^{208}Pb [21], ^{17}F + ^{58}Ni [22], ^{17}F + ^{1}H [23], ^{8}B + ^{28}Si [24], ^{7}Be + ^{58}Ni [25], ^{8}Li + ^{90}Zr [26, 27] and ^{7}Be + ^{28}Si [28, 29]. More recently, a resonant scattering experiment has been performed for the system ^{15}O + ^{4}He [30] and also the capabilities of the facility to be employed as a separator for heavy-ion fusion reaction products were tested [31].

In November 2015 a first experiment of astrophysical relevance was performed. The goal of the measurement was the investigation of the reaction ^{7}Be(n,α)^{4}He by means of the Trojan Horse Method [32, 33] and the surrogate reaction ^{7}Be(d,αp)^{4}He. This reaction is the second most dominant ^{7}Be destruction mechanism in the Big-Bang Nucleosynthesis (BBN) model and might provide a nuclear physics explanation for the ^{7}Li cosmological problem, i.e. the experimental evidence that in metal-poor stars the ^{7}Li abundance is about a factor of 3 smaller with respect to standard BBN theoretical predictions [34]. The data analysis is currently in progress and very promising results have already been communicated to the International Conference Nuclear Physics in Astrophysics VIII (NPA 8, Catania June 2017) [35].

5 Summary and Perspectives

The continuously growing availability of RIBs will make possible the study on earth of nuclear reactions typically occurring only in explosive astrophysical scenarios. The construction of new facilities and the complementarity of the two main production techniques, In-Flight and ISOL, will even widen our present perspectives. Within this framework, a first generation RIB facility was built at INFN-LNL and, after an initial phase mainly devoted to reaction dynamics studies at near-barrier energies, a new program aimed at investigating reactions of astrophysical relevance has recently started.

References

[1] G. Fea, Il Nuovo Cimento **12**, 368 (1935).
[2] Yu.Ts. Oganessian and V.K. Utyonkov, Rep. Prog. Phys. **78**, 03601 (2015).

[3] https://people.nscl.msu.edu/ thoennes/isotopes/
[4] M.S. Smith and K.E. Rehm, Annu. Rev. Nucl. Part. Sci. **51**, 91 (2001).
[5] http://www.nishina.riken.jp/RIBF/
[6] http://isolde.web.cern.ch/
[7] http://www.triumf.ca/research-program/research-facilities/isac-facilities
[8] https://www-win.gsi.de/frs/index.htm
[9] http://pro.ganil-spiral2.eu/laboratory/ganil-accelerators/spiral
[10] http://www.fair-center.eu/
[11] http://pro.ganil-spiral2.eu/spiral2/what-is-spiral2
[12] https://web.infn.it/spes/
[13] https://frib.msu.edu/
[14] V.Z. Maidikov et al., Nucl. Phys. A **746**, 389c (2004).
[15] D. Pierroutsakou et al., Eur. Phys. J. Spec. Top. **150**, 47 (2007).
[16] F.Farinon et al., Nucl. Instrum.Methods B **266**, 4097 (2008).
[17] M. Mazzocco et al., Nucl. Instrum.Methods B **266**, 4665 (2008).
[18] M. Mazzocco et al., Nucl. Instrum.Methods B **317**, 223 (2013).
[19] E. Strano et al., Nucl. Instrum. Methods B **317**, 657 (2013).
[20] D. Pierroutsakou et al., Nucl. Instrum. Methods A **834**, 46 (2016).
[21] C. Signorini et al., Eur. Phys. J. A **44**, 63 (2010)
[22] M. Mazzocco et al., Phys. Rev. C **82**, 054604 (2010)
[23] N. Patronis et al., Phys. Rev. C **85**, 024609 (2012)
[24] A. Pakou et al., Phys. Rev. C **87**, 014619 (2013)
[25] M. Mazzocco et al., Phys. Rev. C **92**, 024615 (2015)
[26] A. Pakou at al., Eur. Phys. J. A **51**, 55 (2015)
[27] A. Pakou at al., Eur. Phys. J. A **51**, 90 (2015)
[28] O. Sgouros et al., , Phys. Rev. C **94**, 044623 (2016)
[29] O. Sgouros et al., , Phys. Rev. C **95**, 054609 (2017)
[30] D. Torresi et al., Phys. Rev. C **96**, 044317 (2017)
[31] E. Strano et al., Nucl. Instrum. Methods A **877**, 293 (2018)
[32] C. Spitaleri et al., Physics of At. Nuclei **74**, 1725 (2011)
[33] R.E. Tribble et al., Rep. Prog. Phys. **77**, 106901 (2014)
[34] C. Broggini et al., Jour. Cosm. Astrop. Phys. **6**, 30 (2012)
[35] L. Lamia et al., EPJ Web of Conferences **165**, 01032 (2017)

Essentials of the macroscopic-microscopic folded-Yukawa approach and examples of its record in providing nuclear-structure data for simulations

Peter Möller[1,*]

[1] *P. Moller Scientific Computing and Graphics, Inc., P. O. Box 1440, Los Alamos, NM 87544, USA*

Abstract. The macroscopic-microscopic model based on the folded-Yukawa single-particle potential and a "finite-range" macroscopic model is probably the approach that has provided the most reliable predictions of a *large* number of nuclear-structure properties for *all* nuclei between the proton and neutron drip lines. I will describe some basic features of the model and the development philosophy that may be the reason for its success. Examples of quantities modeled within the same model framework are, nuclear masses, ground-state level structure, including spins, ground-state shapes, fission barriers, heavy-ion fusion barriers, sub-barrier fusion cross sections, β-decay half-lives and delayed neutron emission probabilities, shape coexistence, and α-decay Q_α energies to name a few. I will show how well it predicted various properties measured *after* published results. Rather than giving an incomplete model description here I will give a timeline of model development and provide references to typical applications and references that are sufficiently complete that several individuals have written computer codes based on these references, codes whose results have excellent agreement with ours.

1 Introduction

Most insight into and understanding of nuclear properties have historically and are currently obtained in terms of surprisingly simple models. In systems which depend mainly on electron behavior their properties can be obtained from solving the Schödinger equation with realistic, known Coulomb potentials. For the nuclear system with its up to a few hundred nucleons a many-body a Scrödinger equation with a potential based on the much more complicated nuclear forces cannot be solved in practice. Therefore other types of models are used. They are often referred to in a condescending manner as phenomenological, but, in a sense all "models" are phenomenological. We will here focus on the remarkable insights into nuclear properties that have been obtained by the "liquid-drop model", the single-particle model and a combination of those, the "macroscopic-microscopic" method that still today is providing much insight about the properties of nuclei.

2 The liquid-drop model

The liquid-drop model has its origin in the semi-empirical mass model, usually attributed to von Weizäcker [1] and Bethe and Bacher [2]. In the liquid drop model the nuclear binding energy, which

*e-mail: mollerinla@gmail.com

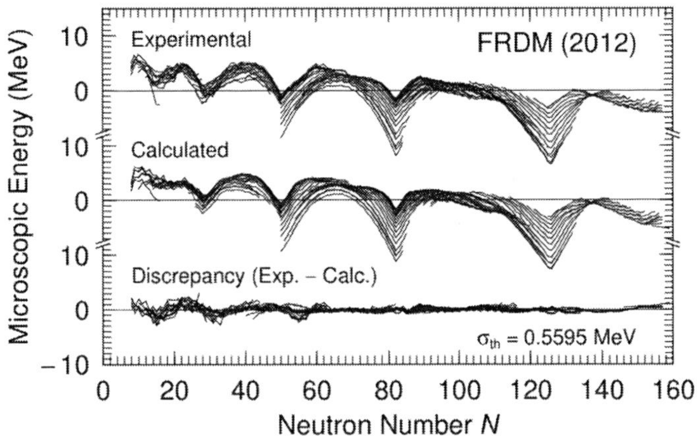

Figure 1. Comparison of experimental and calculated microscopic energies E_{mic} for 2149 nuclei, for a macroscopic model corresponding to the FRDM. The bottom part showing the difference between these two quantities is equivalent to the difference between measured and calculated ground-state masses. There are almost no systematic errors remaining for nuclei with $N \leq 65$, for which region the error is only 0.355 MeV. The results shown in this figure represent our new mass model. The lines are drawn through isotope chains. From Ref. [6].

governs nuclear stability with respect to β and α decay, is the sum of a volume energy term, due to short-range "nuclear" forces, a Coulomb energy term due to the repulsive force between protons, a surface tension term, a symmetry energy term with its origin in the Pauli principle, and finally a pairing term.

$$B(N, Z) = +a_v A - a_s A^{2/3} - a_C \frac{Z^2}{A^{1/3}} - a_I \frac{(N-Z)^2}{A} - \delta(A) \tag{1}$$

This simple expression gives the nuclear binding energy, which for a heavy nucleus is roughly 1600 MeV (8 MeV per nucleon) to an accuracy of one percent. The terms represent in order volume energy, surface energy, Coulomb energy, symmetry energy and a pairing correction. The notation Z, N, and A stand for proton number, neutron number and total nucleon number, respectively. It was highly useful in interpreting the decay chains following element transmutations. However, in some experiments in which uranium was bombarded with neutrons a confusing number of radioactive decay periods, which were difficult to reconcile with properties given by the semi-empirical mass model, were observed. This was all explained when Hahn and Strassmann [3] identified barium in the products following neutron irradiation of uranium and Meitner and Frisch [4] suggested one could think of the nucleus as a deformable charged liquid drop with a surface tension that had split into two smaller nuclei of about equal size in fission. Arguments based on this simple model also showed that the kinetic energies of the two fragments would be very large and Frisch within weeks of the Hahn and Strassmann experiment submitted a paper to Nature [5] in which his measurements confirmed the expected high kinetic energies, thus giving independent confirmation of the fission process. It then turned out that the semi-empirical mass model could be applied to describe this division and Bohr and Wheeler just a few months later suggested the following generalization [7]

$$B(N, Z) = +a_v A - a_s A^{2/3} B_s(\alpha) - a_C \frac{Z^2}{A^{1/3}} B_C(\alpha) - a_I \frac{(N-Z)^2}{A} - \delta(A) \tag{2}$$

Only the Coulomb and surface energies depend on deformation. $B_s(\alpha)$ and $B_C(\alpha)$ are the ratios of the surface and Coulomb energies at deformation α to that for spherical shape. With this simple model one can show that stability with respect to fission decreases with increasing proton number and is completely lost at proton numbers somewhat above $Z = 100$.

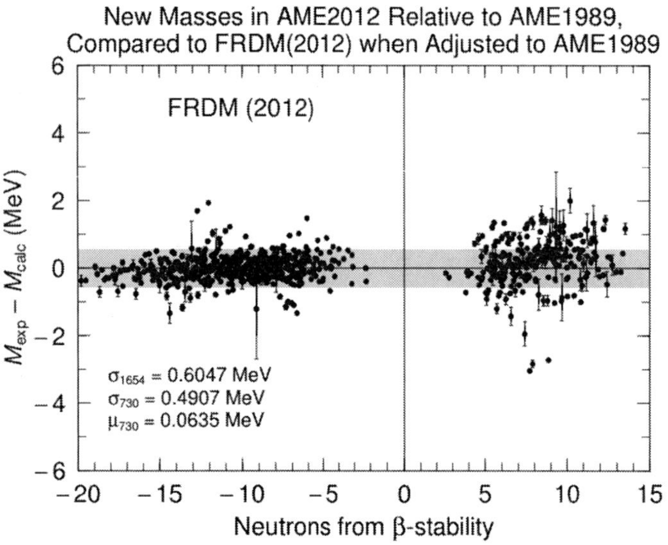

Figure 2. We have adjusted our current mass model (with all the improvements discussed in [6]) to the older AME1989 experimental evaluation to test the extrapability of the model. It agrees better with the AME1989 data base than FRDM(1992), due to improvements in the calculations, 0.6047 MeV versus 0.669 MeV for the previous FRDM(1992). But it also extrapolates much better with only 0.4948 MeV deviations for the new nuclei, versus 0.5817 for the previous FRDM(1992)[8]. From Ref. [6].

3 Single-particle model

Already Bethe and Bacher [2] considered the possibility that microscopic effects might give rise to deviations from the semi-empirical mass model. They noted that in an oscillator single-particle potential there are large gaps in the corresponding single-particle level spectrum and investigated if there were unusually large deviations between their theory and measured masses at $Z = 20$, $N = 20$, namely ^{40}Ca. They found none, but concluded that the masses were not measured sufficiently accurately to observe the effect they expected. In Fig. 1 we compare in the top panel masses calculated in a modern liquid-drop model [6] to measured masses [9]. We observe that there are no particularly large deviations at these nucleon numbers. But for heavier nuclei there are large deviations at the by now well-known magic numbers. Around 1950 a modern single-particle model was developed [10–12]. It showed large gaps at the nucleon numbers where there were large deviations between the semi-empirical mass model and measured masses and it also agreed with observed ground-state spins near magic numbers. But it soon because clear that for large regions of the nuclear chart ground-state spins were not well explained. This observations together with the experimental observations that nuclei in these regions had large intrinsic quadrupole moments led to the suggestion that these nuclei were not spherical but deformed. Motivated by these developments Nilsson in 1955 generalized the spherical single-particle model to describe levels in deformed nuclei for any prescribed spheroidal shape. A large part of the work was to devise a model specification that could be implemented on computers of the day. Nilsson solved this by introducing coordinate transformations that resulted in matrices of low dimensionality about 10 or so [13]. The model results for ground-state spins and low-lying level structure were shown to agree well with experimental observations [14] over a large region of the nuclear chart. It has been one aim of PM to follow this "Copenhagen-Lund" tradition and select viable models based on their ability to describe nuclear properties across the nuclear chart, and also introduce the aim of universality, that is that many nuclear properties are described within a consistent model framework. Before we list our timeline of studies to provide nuclear properties of useful accuracy for astrophysical modeling we need to introduce the third huge development leading to the macroscopic-microscopic method, the Strutinsky shell correction method.

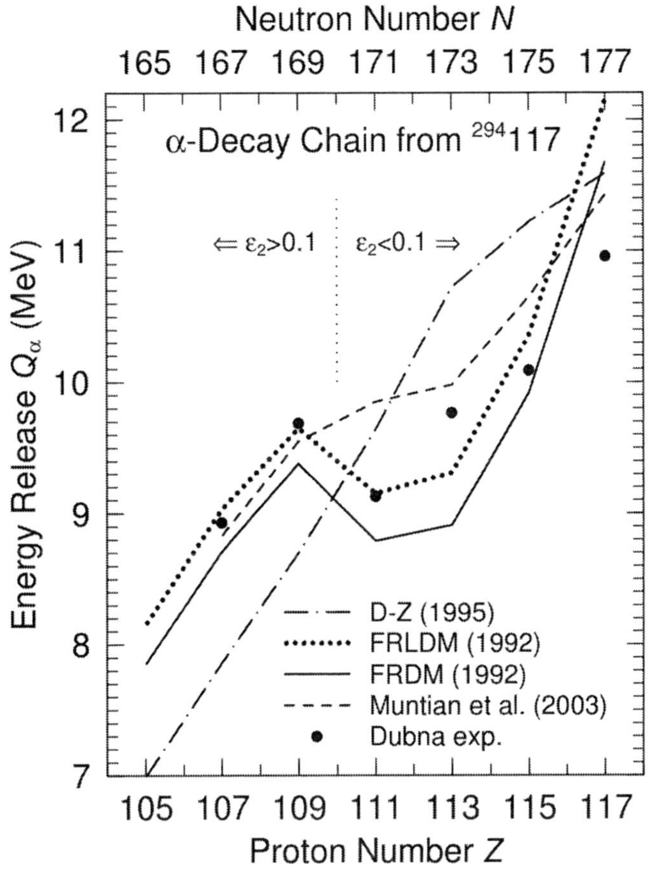

Figure 3. Calculated and measured Q_α -decay energies for ^{294}Ts. The calculated values assume ground-state to ground-state decay, the measured energies may not be, so the differences between measured and calculated values do not directly indicate the degree of (in)accuracy of the underlying mass model. Note that in two of the models the kink observed in the experimental data at $Z = 107$–109 is well reproduced. In this case the kink is related to a change in deformation from spherical to deformed, rather than a gap in the level spectrum at a fixed deformation. D-Z is from Ref. [17] and Muntian from Ref. [18].

4 Strutinsky shell-correction method

Strutinsky introduced a method that combined physics from the macroscopic liquid-drop model and the single-particle model [15, 16]. Briefly, the potential energy for any prescribed deformation is calculated in the following manner. The energy for the charged liquid drop is calculated for this shape, somewhat simplified it is the sum of a surface and electrostatic Coulomb energy. The single-particle energies are calculated. A shell-plus-pairing correction is then obtained from the calculated single-particle spectrum. In Strutinsky's method the correction is about −6 MeV when there is a large gap in the spectrum such as in the neutron spectrum at $N = 126$ for ^{208}Pb. Since there is also a gap in the proton levels at $Z = 82$ the total shell+plus-pairing correction at spherical shape is about −12 MeV. In situations where there is a high density of levels the shell-plus-pairing correction sum is larger than 10 MeV. We now give some examples of how we have used the macroscopic-microscopic method to provide a multitude of nuclear properties across the nuclear chart. With the continuing increase of computer power over decades it is only during the past 20 or so years this method could be exploited to its full potential.

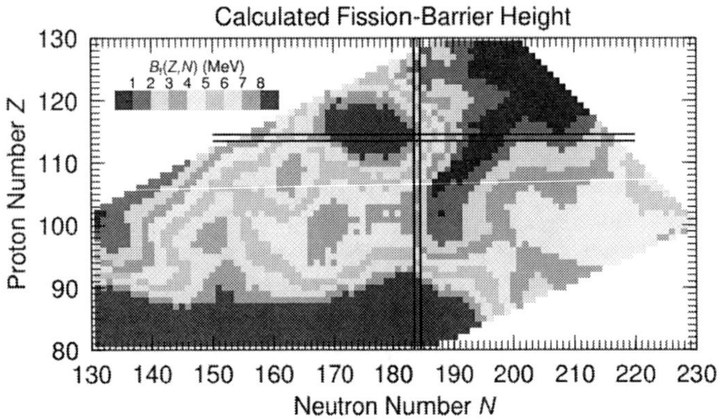

Figure 4. Calculated fission-barrier heights for 3282 nuclei. The highly variable structure is mostly due to ground-state shell effects. Ground-state shell effects are particularly strong in the deformed regions around $^{252}_{100}Fm_{152}$, $^{270}_{108}Hs_{162}$ and in the nearly spherical region near the next doubly magic nuclide postulated to be at $^{298}_{114}Fl_{184}$. The strongest shell effects in this calculation are slightly offset to the left with respect to this isotope. From Ref. [27].

5 Applications of the macroscopic-microscopic method

The first development in this sequence was the introduction of the folded-Yukawa single-particle potential in 1972 [19]. The treatment of the spin-orbit force contribution to the potential has a decisive impact on its success. In the discovery paper the authors used a spin-orbit strength determined in Ref. [20]. In 1973 PM and S-G Nilsson did an extensive study of the optimum spin-orbit strength in the folded-Yukawa and decided on values that are still used today [21]. The first macroscopic-microscopic mass model based on these ideas and an improved macroscopic finite-range "liquid-drop" model [22] appeared in 1981 [23, 24]. Unfortunately, in hindsight it only included about 4000 nuclei, not sufficient to include all r-process nuclei or all nuclei of future interest in the superheavy region. The argument at the time was that we should not blindly let the code crank out numbers without further test of the reliability of the theory. Once comparisons with various types of experimental data and increased computer power were available a mass table extending from the proton to the neutron drip line was completed [8]. This investigation provided masses based on two different macroscopic models, the finite-range liquid-drop model and the finite-range droplet model. It turned out that the latter significantly improved mass predictions. We also note that the expressions used for the neutron and proton single-particle potential radius and depths were since 1972 [19] based on the droplet model ideas and subsequently found to be very realistic and essential [6]. Further development and improved computer performance led to the current mass model FRDM(2012) [6]. As several other groups have done by now, implementation of the FRDM/FRLDM macroscopic-microscopic model can be done based on the complete specifications and mathematical detail found in Refs [6, 8, 19, 22, 25, 26]. It might be useful to note that in Ref. [19] some misprints in Ref. [25] are enumerated.

With this model we also obtain ground-state multipole moments, or equivalently nuclear ground-state shapes. Special types of rare shapes are axially asymmetric and reflection asymmetric shapes. Calculated occurrences of this type of shapes are discussed in Refs. [28, 29].

For some nuclei the potential-energy function versus shape can have several minima. One example is the fission isomeric states which minima correspond to a nuclear shape elongation with the major and minor axes having a ratio of 2:1. Another situation is when there are two minima: one oblate and one prolate, but none as deformed as the fission isomeric state. This situation is referred to as shape coexistence. We have studied globally shape coexistence in Refs. [31, 32].

However, of most interest is perhaps not how well models agree with known data to which its fitted parameters were adjusted. Very important is how reliably properties can be predicted. In Refs. [33, 34] it was shown that the differences between the previously published masses [8] and several hundred new measurements were substantially smaller than in the region to which the model param-

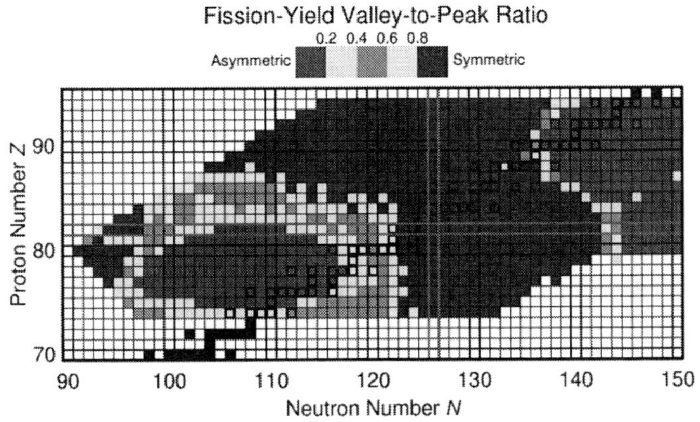

Figure 5. Calculated symmetric-yield to peak-yield ratios for 987 fissioning systems. Black squares (open in colored regions, filled outside) indicate β-stable nuclei. We find a new, contiguous region of asymmetric fission separated from the classical location of asymmetric fission in the actinides by an extended area of symmetric fission. From Ref. [30].

eters were adjusted. To obtain similar insight we have adjusted the parameters of the FRDM(2012) to a more limited mass data set [35] and studied the calculated masses obtained for masses measured subsequently. The agreement with subsequently measured masses is shown in Fig. 2 and shows encouraging predictive capability. Another way to assess the predictive properties is to compare do superheavy element (SHE) properties. Some of their properties are obtained through their α-decay chains. In Fig. 3 we show a comparison of the FRDM(1992) predictions to more recently observed [36] α-decay Q values of the decay of ^{294}Ts. It is clear the predictions well reproduce the observed interesting structure in the decay chain.

In fission, as we remarked above, the nucleus evolves from a single (ground-state) shape to two separated fission fragments. This process can be studied in the macroscopic-microscopic model by studying the energy of nuclear shapes involved in the transition between the ground state and separated fragments. A study of a few selected shapes (the only possibility at that time with the limited computer power) showed a definite signature of asymmetric fission for actinides but symmetric fission for somewhat lighter nuclei [37, 38]. It was much later that the macroscopic-microscopic method could be exploited by calculating the potential energy between the ground state and separated fragments for the selection of shapes that should be considered, namely elongation, neck radius, left and right fragment spheroidal deformation, and mass split between the fragments [39]. A poorly appreciated aspect in this type of calculations is: what is required to identify correctly fission saddle points? A more in-depth account of these aspects is in Ref. [40]. Calculated fission barrier-heights based on these methods are in Ref. [27]. We show in Fig. 4 3282 barrier heights calculated in this approach.

For many simulation applications it is of interest to have available the fission-fragment mass distributions. Recently we showed that a Brownian shape motion on calculated five-dimensional potential-energy surfaces well reproduced the observed fission-fragment mass distributions [41–43].

Encouraged by these studies which showed very good agreement with experimental fission fragment mass distributions, we went ahead, as we usually do, and made some predictions, so that our model could be tested, of fission fragment mass distributions for an extended region of 987 heavy nuclei [30]. We found a new contiguous region of nuclei predicted to undergo asymmetric fission, see Fig. 5. An earlier "discovery" experiment [44] (which motivated our study) had shown that ^{180}Hg fissioning asymmetrically in β-delayed fission. There are still only scattered experiments of fragment distributions in this region, mainly in the transition regions between symmetric and asymmetric fission, so the predictions still await serious experimental studies. Currently in progress are calculations of fission yields of r-process nuclei for use in r-process fission recycling modeling.

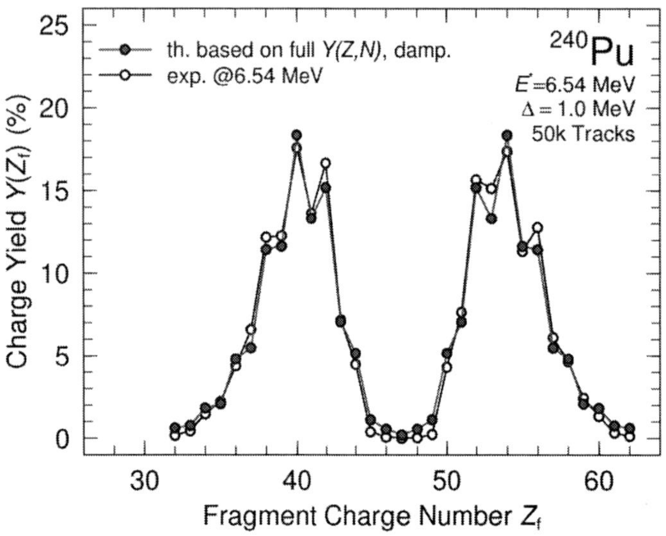

Figure 6. Calculated fission-fragment charge yields in thermal neutron-induced fission of ^{239}Pu. The calculated yield is based on a calculation of $Y(Z, N)$ in a recent extension of the Brownian shape-motion model [45] which is then summed over neutron number.

Recently we have generalized the Brownian shape-motion model so that we now model the fission-fragment yields $Y(Z, N)$. The modification is not so much in the Brownian shape-motion model but rather in devising a method to calculate the potential energy of the fissioning nucleus for different Z/N ratios in the two emerging fragments. This leads to two shape asymmetry variables one in Z and one in N. The method and benchmarks are presented in Ref. [45]. In Fig. 6 we compare calculated and experimental charge distributions of ^{240}Pu. To obtain the charge distribution we summed for each Z the calculated $Y(Z, N)$ over neutron number N.

Another substantial modeling effort has been and is to calculate β-decay properties, specifically decay half-lives and delayed neutron-emission probabilities. Our first global results are in Ref. [33]. A next generation, improved calculation and release will be available in the first half of 2018. PM maintains a web page with publications, figures appearing in these publications and interactive features at **URL: http://t2.lanl.gov/nis/molleretal/**.

References

[1] C. F. von Weizäcker, Z. Phys. **96** (1935) 431
[2] H. A. Bethe and R. F. Bacher, Rev. Mod. Phys. **8** (1936) 82
[3] O. Hahn and F. Strassmann, Naturwiss. **27** (1939) 11
[4] L. Meitner and O. R. Frisch, Nature **143** (1939) 239
[5] O. R. Frisch, Nature **143** (1939) 276
[6] P. Möller, A. J. Sierk, T. Ichikawa, and H. Sagawa, Atomic Data and Nuclear Data Tables **109–110** (2016) 1
[7] N. Bohr and J. A. Wheeler, Phys. Rev. **56** (1939) 426
[8] P. Möller, J. R. Nix, W. D. Myers, and W. J. Swiatecki, Atomic Data Nucl. Data Tables **59** (1995) 185
[9] M. Wang, G. Audi, A. H. Wapstra, F. G. Kondev, M. MacCormick, X. Xu, and B. Pfeiffer, Chin. Phys. **C36** (2012) 1603
[10] M. Mayer, Phys. Rev. **75** (1949) 1969
[11] M. Mayer, Phys. Rev. **78** (1950) 16

[12] M. Mayer, Phys. Rev. **78** (1949) 22
[13] S. G. Nilsson, Kgl. Danske Videnskab. Selskab. Mat.-Fys. Medd. **29**:No. 16 (1955)
[14] B. R. Mottelson and S. G. Nilsson, Kgl. Danske Videnskab. Selskab. Mat.-Fys. Skr. . **1**:No. 8 (1959)
[15] V. M. Strutinsky, Nucl. Phys. **A95** (1967) 420
[16] V. M. Strutinsky, Nucl. Phys. **A122** (1968) 1
[17] J. Duflo and A. P. Zuker, Phys. Rev. C **52** (1995) R23
[18] I. Muntian, Z. Patyk, and A. Sobiczewski, Physics of Atomic Nuclei, **66** (2003) 1015
[19] M. Bolsterli, E. O. Fiset, J. R. Nix, and J. L. Norton, Phys. Rev. C **5** (1972) 1050
[20] W. Ogle, S. Wahlborn, R. Piepenbring, and S. Fredriksson, Rev. Mod. Phys. **43** (1971) 424
[21] P. Möller and J. R. Nix, Proc. Third IAEA Symp. on the physics and chemistry of fission, Rochester, 1973, vol. I (IAEA, Vienna, 1974) p. 103
[22] H. J. Krappe, J. R. Nix, and A. J. Sierk, Phys. Rev. C **20** (1979) 992
[23] P. Möller and J. R. Nix, Nucl. Phys. **A361** (1981) 117
[24] P. Möller and J. R. Nix, Atomic Data Nucl. Data Tables **26** (1981) 165
[25] J. Damgaard, H. C. Pauli, V. V. Pashkevich, and V. M. Strutinsky, Nucl. Phys. **A135** (1969) 432
[26] K. T. R. Davies, A. J. Sierk, and J. R. Nix, Phys. Rev. C **13** (1976) 2385
[27] P. Möller, A. J. Sierk, T. Ichikawa, A. Iwamoto, and M. Mumpower, Phys. Rev. C **91** (2015) 024310
[28] P. Möller, R. Bengtsson, B. G. Carlsson, P. Olivius, and T. Ichikawa, Phys. Rev. Lett. **97** (2006) 162502
[29] P. Möller, R. Bengtsson, B. G. Carlsson, P. Olivius, T. Ichikawa, H. Sagawa, and A. Iwamoto Atomic Data and Nuclear Data Tables **94** (2008) 758
[30] P. Möller and J. Randrup, Phys. Rev. C **91** (2015) 044316
[31] P. Möller, A. J. Sierk, R. Bengtsson, H. Sagawa, and T. Ichikawa, Phys. Rev. Lett. **103** (2009) 212501
[32] P. Möller, A. J. Sierk, R. Bengtssson, H. Sagawa, and T. Ichikawa, Atomic Data and Nuclear Data Tables, **98** (2012) 149
[33] P. Möller, J. R. Nix, and K.-L. Kratz, Atomic Data Nucl. Data Tables **66** (1997) 131
[34] P. Möller, R. Bengtsson, K.-L. Kratz, and H. Sagawa, Proc. International Conference on Nuclear Data and Technology, April 22–27, 2007, Nice, France, (EDP Sciences, (2008) p. 69, ISBN 978-2-7598-0090-2), and http://t2.lanl.gov/nis/molleretal/publications/nd2007.html
[35] G. Audi, Midstream atomic mass evaluation, private communication (1989), with four revisions
[36] Yu. Ts. Oganessian, J. Phys. G: Nucl. Part. Phys. **34** (2007) R165
[37] P. Möller and S. G. Nilsson, Phys. Lett. **31B** (1970) 283
[38] P. Möller, Nucl. Phys. **A192** (1972) 529
[39] P. Möller, D. G. Madland, A. J. Sierk, and A. Iwamoto, Nature **409** (2001) 785
[40] P. Möller, A. J. Sierk, T. Ichikawa, A. Iwamoto, R. Bengtsson, H. Uhrenholt, and S. Åberg, Phys. Rev. C **79** (2009) 064304
[41] J. Randrup and P. Möller, Phys. Rev. Lett. **106** (2011) 132503
[42] J. Randrup, P. Möller, and A. J. Sierk, Phys. Rev. C **84** (2011) 034613
[43] J. Randrup and P. Möller, Phys. Rev. C **88** (2013) 064606
[44] A.N. Andreyev, J. Elseviers, M. Huyse, P. Van Duppen, S. Antalic, A. Barzakh, N. Bree, T.E. Cocolios, V.F. Comas, J. Diriken et al., Phys. Rev. Lett. **105**, 252502 (2010)
[45] P. Möller and T. Ichikawa, Eur. Phys. J. A **51** (2015) 173

ANC experiments for nuclear astrophysics in NPI CAS

J. Mrazek[1,★], *V. Burjan*[1], *V. Kroha*[1], *A.M.Mukhamedzhanov*[3], *R.Tribble*[3], *C. Spitaleri*[2], *I. Siváček*[1], *V. Glagolev*[1], *Š. Piskoř*[1], *G.R. Pizzone*[2], *M. La Cognata*[2], *G. D'Agata*[2], *A. Tumino*[2], *L. Lamia*[2], *R. Sparta*[2], *G. Rapisarda*[2], and *S. Romano*[2]

[1]*Nuclear Physics Institute of the Czech Academy of Sciences, p.r.i., Řež 130, 25068 Řež, Czech Republic*
[2]*Instituto Nazionale di Fisica Nucleare, LNS, Via S. Sofia 62, 95125 Catania CT, Italy*
[3]*Cyclotron Institute, Texas A&M University, College Station, Texas 77842*

Abstract. Asymptotic Normalization Coefficients method is one of the indirect methods used in nuclear astrophysics. The method allows to deduce a direct part of a radiative capture cross section from the measurement of a two body reaction under specific conditions. The experimental works started in collaboration of NPI CAS, TAMU and INFN LNS from 90-ties and continue till the present. The introduction to this method is presented and experimental experiences from NPI CAS and the collaboration are shared.

1 Introduction

Nuclear reactions at low energies are of a specific interest in astrophysics. A number of scenarios of stellar nucleosynthesis involve low energy reactions, frequently radiative capture and these reaction frequently - due to the low energy and other conditions proceed peripherally. However, direct measurement of these reactions is difficult and technological development makes it more and more accessible only in the last decades.

Indirect methods can provide the information that is currently not possible to obtain directly, but also can bring another piece of information to the direct measurements, that is necessary to draw proper conclusions.

Asymptotic Normalization Coefficients method can extract information about direct radiative capture like (p,γ), (n,γ), (α,γ) at low energies. There is also one auxiliary advantage for older accelerator devices, who were not modernized or decommissioned in the last decades. The older accelerator types can be used with indirect methods, if they can accelerate light ion beams like deuteron and ^{3}He at energies around 20 MeV, which is also a case of the NPI CAS cyclotron U120M. The ANC method measurements were performed here from the dawn of the method from 90-ties until now in collaboration NPI - TA&MU - INFN-LNS developments driven by A.M.Mukhamedzhanov, R.Tribble and V.Kroha.

1.1 Historical excursion

When nuclear reactions started to be studied, Fermi [1] discovered in 1934 an enhancement (resonant) of the cross sections. Soon Breit and Wigner [2] presented their theory of resonances and later,

★e-mail: mrazek@ujf.cas.cz

the hypothesis of independence was provided by Bohr 1936, where the reaction proceeds through a compound state and the nucleus looses information about the creation channel. Ghoshal et al. [3] have shown a convincing experimental evidence for such a behavior.γ and β spectroscopy gave the pictures of nuclear structure, and showed an importance of a collective excitations.

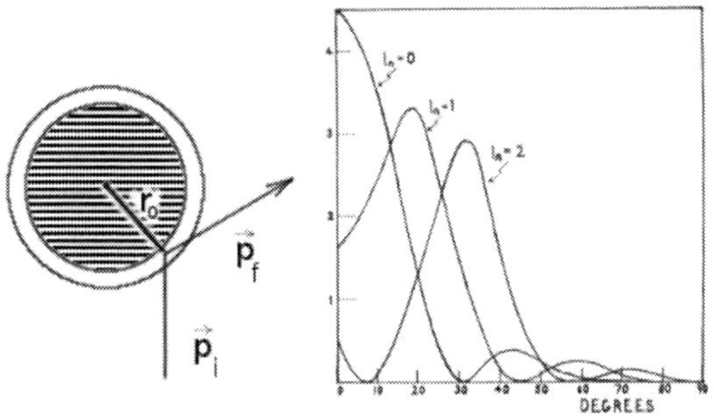

Figure 1. Left panel: S.T.Butler had the idea that the nucleus contains a black (absorption) core and a grazing region, where the direct reactions occurs [4]. Right panel: Angular distributions of direct reactions peak at certain angle that depends on the transferred momentum $L = |\vec{r_0} \times (\vec{p_f} - \vec{p_i})|$. [5]

Single-particle transfer reactions started to be studied in 50-ties and seemed to be they are important for understanding single particle excitations (collective excitations studies were developed with β and γ spectroscopy). S.T.Butler [4, 5] pointed out the validity of the Born approximation and he made a great contribution to direct reactions as a tool for nuclear structure (see Fig.1).

1.2 Optical model

A description of a nuclear reaction can be written [6] in a very general way using the the Schrödinger equation $(H_x + H_A + T + \Sigma_{i\in x, j\in A} V_{ij})\Psi_\alpha = E\Psi_\alpha$ where all nucleons of a projectile x and a target A mutually interact and the index α stands for the input channel or a partition $x + A$.

It is useful to accept some simplifications. The shell model suggests to describe a nucleus as nucleons in a central potential U, that represents an average field of all nucleons with corrections that are taken into account as residual interactions V_{ij}. By this way, it is intuitive, that the interaction of the approaching projectile and the nucleus is (at large distances) driven by this potential U. If there is no exchange of energy or nucleons, then this potential U should solely describe the interaction - the elastic scattering. The Schrödinger equation of the system may be written as $(H + T + U)\Psi = E\Psi$. Then, a direct reaction may be taken as a perturbation of the elastic scattering [6].

Under these considerations, the Schrödinger equation is separable into nuclear coordinates and relative coordinates of the projectile. The solution Ψ_α can be written as a product of nuclear Φ_α and relative $\chi_\alpha(r)$ wave functions $\Psi_\alpha = \Phi_\alpha\, \chi_\alpha(r)$, where the corresponding Schrödinger equation may be written $(T + U)\,\chi_\alpha(r) = E\,\chi_\alpha(r)$.

The most traditional shape of the optical potential U is a Woods-Saxon form. To describe the loss of flux due to reactions the potential must also have a complex term and to describe the experimentally observed orbitals it also has an LS term, and frequently a surface term is also introduced.

A transfer reaction has a different output channel, partition $Y + B$ named β, so it is possible to find another optical potential and a relative wave function $U_\beta, \chi_\beta(r)$.

1.3 DWBA approach

A typical transfer reaction may be depicted as in the diagram in Fig.2. Using the above considerations about the optical potential it is possible to deduce a very important tool in direct nuclear reactions - the Distorted-Wave Born Approximation. The details can be found in textbooks e.g. by Glendening

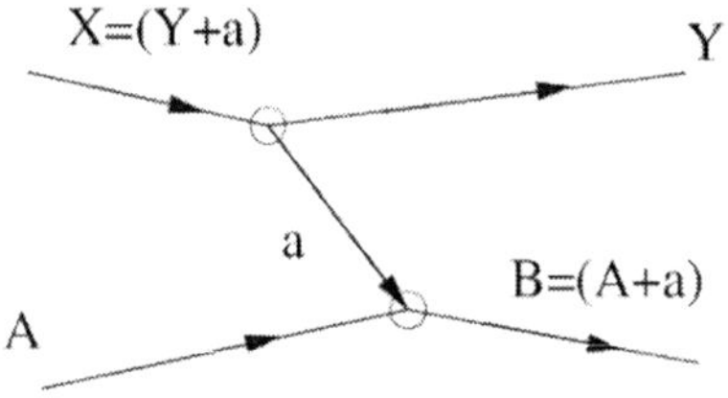

Figure 2. Diagram of a transfer reaction. The vertex contains the information for direct radiative capture.

[6] in chapter 5. This (DWBA) approximation leads to a formula for a transition amplitude between the states before and after the reaction.

The matrix element of the transition between (inelastic) incoming and outgoing channels can be written (in post-form) as

$$M(E_i, \theta_{c.m.}) = \sum_{Ma} \langle \chi_f^{(-)} I_{Aa}^B |\, \Delta V \,| I_{Ya}^X \chi_i^{(+)} \rangle, \tag{1}$$

where I_{Aa}^B is the radial overlap function of the system $A + a$ and the nucleus B, χ_i and χ_f are incoming and outgoing (by the potential U) distorted waves, ΔV is the perturbation part of the potential $\Delta V = V_\beta - U_\beta$, where V is a full (unknown to us) potential and U is the optical potential.

The radial overlap function - in general I_{ab}^c can be approximated by (model calculated using e.g. Woods-Saxon potential well) bound state wave functions $\varphi_{n,l_c,j_c}(r_{ab})$ through a factor S_{ab}

$$I^c_{ab\, l_c j_c}(r_{ab}) = S^{1/2}_{ab\, l_c j_c} \varphi_{n_c l_c j_c}(r_{ab}). \tag{2}$$

The factor $S^{1/2}_{ab\, l_c j_c}$ is called the spectroscopic factor (SF) and it was intensively used to study the occupation of different nuclear orbitals. Nowadays, it is still a very good tool to inspect the nuclear structure, but time has shown that it has certain limitations. The SF's depend on the selection of the optical model (OM) potential and there are in general several ambiguities concerning OM's: i) several different potential families can reproduce the angular distribution data, ii) the cross section scales with the potential radius r_0, iii) the (Woods-Saxon) depth is usually tuned to reproduce the bound states (well depth prescription).

Another important fact about SFs was discussed recently by several authors [7–9]: SFs rely on internal wave functions, which implies that in fact they are not observables.

Despite these facts, it is possible to face these difficulties by a careful treatment of the analysis as was shown e.g. by X.Liu [10]. He compared the previous works with (d,p) and (p,d) reactions with ^{12}C as a target. He finds differences of factor 5 and by a consistent treatment he could reach 15% variation of the reevaluated data.

The experimentally measured angular distributions $\frac{d\sigma}{d\Omega}$ can be reproduced by DWBA calculations.

$$\frac{d\sigma}{d\Omega} = \Sigma_{j_B j_x} S_{Aal_B j_B}\, S_{Yal_x j_x}\, \sigma^{DW}_{l_B j_B l_x j_x} \tag{3}$$

$\frac{d\sigma}{d\Omega}$ and $\sigma^{DW}_{l_B j_B l_x j_x}$ are related through a coefficient that basically consists of two SF, one of them is usually (for light particles like d, ^{3}He) known, so the analysis will offer the spectroscopic factor of the studied state.

2 Asymptotic Normalization Coefficients

Mukhamedzhanov and Timofeyuk have pointed out in 1990 [11] that the direct radiative capture cross sections of low energy astrophysical reactions usually contain the same radial overlap integral (I_{ab}^c) as the (direct) transfer reactions. Since these direct captures frequently proceed at large distances from the nucleus, it is possible to use asymptotic expressions for the radial overlap functions (and bound state wavefunctions (see eq.2))

$$\begin{aligned} I^{\alpha}_{\beta\gamma l_\alpha j_\alpha}(r_{\beta\gamma}) &\xrightarrow{r_{\beta\gamma}>R_N} C^{\alpha}_{\beta\gamma l_\alpha j_\alpha}\frac{W_{-\eta_\alpha,l_\alpha+1/2}(2\kappa_{\beta\gamma}r_{\beta\gamma})}{r_{\beta\gamma}} \\ \varphi_{n_\alpha l_\alpha j_\alpha}(r_{\beta\gamma}) &\xrightarrow{r_{\beta\gamma}>R_N} b_{\beta\gamma l_\alpha j_\alpha}\frac{W_{-\eta_\alpha,l_\alpha+1/2}(2\kappa_{\beta\gamma}r_{\beta\gamma})}{r_{\beta\gamma}}, \end{aligned} \tag{4}$$

where $W_{-\eta_\alpha,l_{\alpha+1/2}}$ is a Whitaker function that has a known asymptotics, $W_{-\eta_I,l+\frac{1}{2}}(-2k_I r) \rightarrow e^{-k_I r+\eta_I ln(2k_I r)}$.

The coeficients C^2 and b^2 are called ANC and single particle ANC (SPANC), respectively. If we use the above approximation with the eq.3, we obtain a modified formula for comparison of experimental data with the DWBA model:

$$\frac{d\sigma}{d\Omega} = \Sigma_{j_B j_x}(C^B_{Aal_B j_B})^2(C^X_{Yal_x j_x})^2\frac{\sigma^{DWBA}_{l_B j_B l_x j_x}}{b^2_{Aal_B j_B}b^2_{Yal_x j_x}} = \Sigma_{j_B j_x}(C^B_{Aal_B j_B})^2(C^X_{Yal_x j_x})^2 R_{l_B j_B l_x j_x} \tag{5}$$

The traditional method of finding spectroscopic factors turns into finding ANCs (for light projectiles can be C^X_{Ya} known) based on DWBA analysis giving the function R. At first, it does not seem to be an improvement, but in fact, the ANCs have several advantages over SF as was already mentioned in sec. 1.3.

- for a peripheral reaction, the ANCs have at most a small dependence on the selection of the potential. $R_{l_B j_B l_x j_x}$ from eq.5 is nearly independent of b^2
- ANCs are defined in the nuclear exterior, so they are "observable"
- ANC is related to the Nuclear Vertex Constant (NVC) $G^{\alpha}_{\beta\gamma l_\alpha j_\alpha} = -e^{i\pi[(l_\alpha+\eta_\alpha)/2]}\frac{\sqrt{\pi}}{\mu_\alpha}C^{\alpha}_{\beta\gamma l_\alpha j_\alpha}$, the same constant that appears [12, 13] in direct radiative capture reaction a+A and where $\sigma \sim G^2_{ABa}$.

2.1 Validity tests

As it was mentioned above, the validity of extraction of ANCs from transfer reactions depends on the peripherality of the transfer reaction. Therefore, it is necessary to perform different checks of the analysis. After the extraction of angular distributions and finding the optical potential, it is useful to make a comparison of the model σ^{DWBA} for different R_{cutoff} parameters for the integration. The comparison should show that the selected optical potential is not much dependent on the nucleus interior. See left panel in Fig. 3, where the reaction p+^{9}Be was studied by A.M.Mukhamedzhanov et al. [14] using the reaction $^9B(^{10}B,^{10}B)^9Be$ at Texas A&M.

Since the single particle radial wave function φ behavior depends quite strongly on the optical model potential, it is possible to inspect the radial behavior of φ varying the potential parameter r_0. The tail should correspond to the Whitaker function. It should be visible, that the single particle wave function reaches the asymptotic form and can be replaced by the Whitaker function even at short distances, e.g. middle panel in Fig.3 at $r > 4.5$ fm [14].

Another important check is the test of dependence of ANC (C^2) on optical potential parameters. An example is given in Fig.3, right panel from A.M.Mukhamedzhanov et al. [14], where the ANC was plotted in dependence on SPANC b, as it represented a better way to determine the changes in r_0

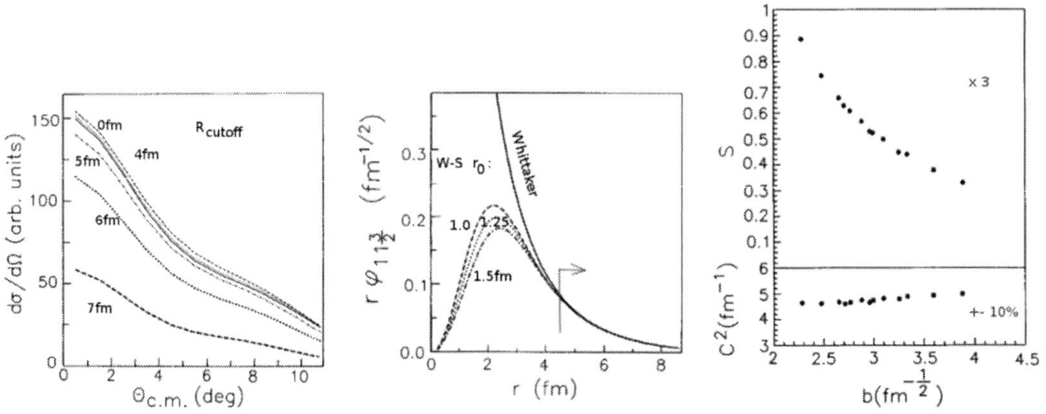

Figure 3. Left panel: The R_{cutoff} parameter was varied from zero to 7 fm and it is shown that the dependence of the σ^{DWBA} is weak up to 4 fm as done by A.M.Mukhamedzhanov et al. [14], Middle panel: The single particle radial wave function φ variation inside the nucleus and the asymptotic above $r > 4.5$ fm, Right panel: ANC varies bellow 10% for range of different optical model parameters, for comparison a strong dependence of the SF is shown.

and a parameters of the bound state potential. The variation of the ANC was there shown to be bellow 9%, while for comparison, SF varied by a factor 3 for the same variation of r_0, a.

With the appearance of resonances near the threshold (or the reaction in study) region, the direct radiative capture is not the only nor the main process contributing to the (p, γ) process. The resonant capture plays an important role and frequently its contribution is much more important. However, the direct capture can contribute in an indirect way. The coherent sum of the direct and indirect contributions can create important effects in the cross section (or astrophysical S factor) in a region of interest.

An instructive example is the $^{15}N(p,\gamma)^{16}O$ reaction measured and analyzed by our group, A.M.Mukhamedzhanov et al. [15], where two 1^- resonances (left panel of Fig.4) dominate the low energy reagion. The deduced ANC in [15] was $C^2 = 192. \pm 26$ fm^{-1} ($C = 13.856$ fm$^{-1/2}$), the deduced astrophysical S factor was $S(0) = 36.0(6.0)$ keVb, which was below the previous findings and this resulted in a change of the estimation of the leak from this CNO cycle reaction from one every 2200 (300) cycles to one per 1200, which is almost a factor two. This measurement was later confirmed by P.J.LeBlanc and LUNA collaboration [16], where the $S(0)$ was found $S(0) = 39.6 \pm 2.6$ keVb.

In response to [16], Mukhamedzhanov, La Cognata and Kroha [17] have discussed the treatment of ANC in the above work, which resulted in a new perspective of using ANCs as an auxiliary information to pinpoint spectroscopic factors.

For a comparison the right panel of Fig.4 shows the influence of different values of ANCs to the description of resonances.

3 Summary

The principles of the indirect method for determination of direct radiative capture - Asymptotic Normalization Coefficients method (ANC) - and it's evolution from spectroscopic factors were presented. To validate the approximations of the method, it is necessary to do specific checks with the model. Examples from real case analysis were presented. The importance of the knowledge of ANCs also

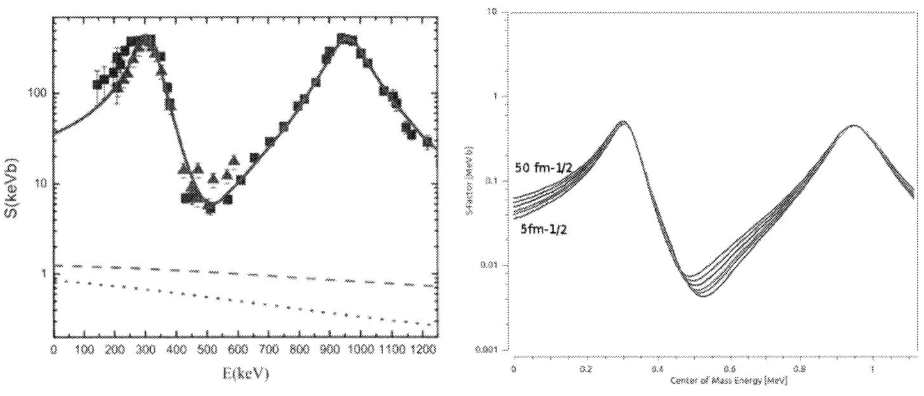

Figure 4. Left panel: Astrophysical S-factor based on R-matrix fit of the known data and the deduced ANC in A.M.Mukhamedzhanov et al. [15]. The dashed line is ANC for the ground state. Right panel: variation of ANC (C_{mean} = 13.85 fm$^{-1/2}$ and 10 fm$^{-1/2}$ in red) entering the R-matrix fit shows the highest sensitivity at very low energy, which is a region for e.g. CNO scenario.

in resonant radiative capture was pointed out in the example. This publication was supported by OP RDE, MEYS, Czech Republic under the project SPIRAL2-CZ, CZ.02.1.01/0.0/0.0/16_013/0001679.

References

[1] E. Fermi, E. Amaldi, O. D'Agostino, F. Rasetti, E. Segre, Proceedings of the Royal Society of London A: Mathematical, Physical and Engineering Sciences **146**, 483 (1934)
[2] G. Breit, E. Wigner, Phys. Rev. **49**, 519 (1936)
[3] S.N. Ghoshal, Phys. Rev. **80**, 939 (1950)
[4] S.T. Butler, Phys. Rev. **106**, 272 (1957)
[5] S.T. Butler, Phys. Rev. **80**, 1095 (1950)
[6] N.K. Glendenning, *Direct Nuclear Reactions* (ACADEMIC PRESS, INC., 1983)
[7] R. Furnstahl, H.W. Hammer, Physics Letters B **531**, 203 (2002)
[8] B.K. Jennings, ArXiv e-prints (2011), 1102.3721
[9] A.M. Mukhamedzhanov, A.S. Kadyrov, Phys. Rev. C **82**, 051601 (2010)
[10] X.D. Liu, M.A. Famiano, W.G. Lynch, M.B. Tsang, J.A. Tostevin, Phys. Rev. C **69**, 064313 (2004)
[11] A.M. Mukhamedzhanov, N.K. Timofeyuk, JETP Lett **52**, 282 (1990)
[12] L.D.Blokhintsev, Fiz.Elem.Chastits At.Yadra **8**, 1189 (1977)
[13] L.D.Blokhintsev, Fiz.Elem.Chastits At.Yadra **15**, 1296 (1984)
[14] A.M. Mukhamedzhanov, H.L. Clark, C.A. Gagliardi, Y.W. Lui, L. Trache, R.E. Tribble, H.M. Xu, X.G. Zhou, V. Burjan, J. Cejpek et al., Phys. Rev. C **56**, 1302 (1997)
[15] A.M. Mukhamedzhanov, P. Bém, V. Burjan, C.A. Gagliardi, V.Z. Goldberg, Z. Hons, M. La Cognata, V. Kroha, J. Mrázek, J. Novák et al., Phys. Rev. C **78**, 015804 (2008)
[16] P.J. LeBlanc, G. Imbriani, J. Görres, M. Junker, R. Azuma, M. Beard, D. Bemmerer, A. Best, C. Broggini, A. Caciolli et al., Phys. Rev. C **82**, 055804 (2010)
[17] A.M. Mukhamedzhanov, M. La Cognata, V. Kroha, Phys. Rev. C **83**, 044604 (2011)

EPJ Web of Conferences **184**, 01015 (2018) https://doi.org/10.1051/epjconf/201818401015
9th European Summer School on Experimental Nuclear Astrophysics

Gas detectors for nuclear physics experiments

Dimitra Pierroutsakou[1,*]

[1] *INFN - Sezione di Napoli, Via Cintia, I-80126 Napoli, Italy*

Abstract. In this lecture I will present the operation principle and the different kinds of gas detecting systems for charged particles employed in high-energy and low-energy physics environments, with particular focus on the requirements of nuclear physics experiments with low-energy Radioactive Ion Beams (RIBs). I will show in more details an example of gas detector used at the RIB in-flight facility EXOTIC, for the ion beam tracking and for time of flight measurements. Finally, I will discuss the use of an active target in nuclear physics experiments with RIBs together with some key improvements of first generation devices required for facing the challenges of more intense RIBs.

1 Operation principle and different kinds of gas detectors

Gas detectors are planned or being constructed at every nuclear physics facility because they have some important advantages with respect to solid state devices for charged particle detection. To summarize their characteristics, they have: a) good stability, robustness and aging compared to solid state detectors; b) high radiation length; c) possibility to tune the effective thickness; d) low energy thresholds; e) three dimensional readout/flexible geometry; f) good space and moderate energy resolution; g) medium rates for particles. Finally, for large-area coverage with low material budget they are still the only solution.

Their **operation principle** is based on the ionization and excitation of the gas molecules caused by the passage of the charged particle. Electron-ion pairs are created either from the particle itself or by δ-rays and the average energy necessary to create one ion-pair, about W=25-30 eV, is not much dependent on the gas. Some non-ionizing energy loss (excitation) makes the number of pairs lower and this is important for the detector energy resolution. The number N of produced pairs follows a Poisson distribution $\sigma = \sqrt{N}$. Without electric field, electron-ion pairs diffuse due to random thermal movement and suffer collisions with gas molecules. In these collisions, recombination of ions to neutrals (for molecular ions), electron attachment in gases with large electron affinity and charge exchange can occur. In case of diffusion of electron-ion pairs in electric field, when high voltage (HV) is applied to the electrodes of the detector, we have the superposition of thermal velocity (random) and drift velocity of electrons/ions along the lines of the electric field. Electrons are much faster than positive ions, by a factor of about 1000. A magnetic field parallel to the electric field will curl up electrons in the transverse (to the direction of motion) plane reducing their transverse diffusion.

By increasing the HV and according to the multiplication of the primary number of created electron-ion pairs (gain), the detector can operate in different regimes: **a) Ionization mode** with full charge collection and no multiplication, gain $\approx$ 1; **b) Proportional mode** with multiplication of the

*e-mail: pierroutsakou@na.infn.it

primary ionization, a signal that is proportional to the ionization, possibility to measure dE/dx, where secondary avalanches need quenching (quenching gas is a neutral gas with a lower ionization potential, added to the main detector gas that quenches the ion multiplication by neutralizing ions of the main chamber gas), gain $\approx 10^4$-10^5; c) **Limited proportional mode** where the created space charge distorts the electric field, we have loss of proportionality and strong photoemission that requires strong quenchers or pulsed HV, gain $\approx 10^8$; **d) Geiger mode** with a signal amplitude independent of energy, massive photoemission, where the full length of the anode wire is affected and the discharge can be stopped by HV cut, gain $> 10^8$.

A detector that works in the first regime is the **ionization chamber**. It consists of two metallic electrodes at distance d covering a volume gas and the application of a HV to the electrodes creates a constant electric field.

In the **proportional counter** (proportional regime) that has a single anode wire in a cylindrical cathode, the created electric field $E \sim \frac{1}{r}$ is very strong near the anode wire and weak for larger distances r. The electron-ion pairs created from the passage of a particle drift in the volume while multiplication of the primary charge occurs only near the anode wire through an avalanche process. Many closely spaced anode wires in the same chamber can each act as an independent proportional counter. This is the **Multiwire Proportional Counter (MWPC)** invented by G. Charpak, CERN 1968, Nobel Prize 1992. The basic configuration of a MWPC is two cathode planes and a wire anode in between. It has many advantages: a very flexible geometry and a large detection area ($\approx$ m^2); possibility to work in magnetic field, with a rate capability $\approx 10^4$ Hz/mm^2; gain $\approx 10^4$-10^5; possibility to perform Particle IDentification (PID) through dE/dx measurement and many well developed position encoding methods: binary readout (identification of the closest anode wire to the avalanche), charge division by using resistive anode wires and analogue readout of the signals of a segmented cathode. By using several layers of MPWC giving x-y position it is possible to perform tracking of the charged particle.

A variant of the MWPC is the **Parallel Plate Avalanche Counter (PPAC)**, preferred when fast timing is more important than energy resolution. The basic configuration of a PPAC is two planar electrodes covering a gas at low pressure, P=7-20 mbar, that yields a uniform intense reduced electric field $\frac{E}{P}$. The avalanche multiplication starts immediately from the primary ionization electrons resulting in a very good time resolution (of the order of hundred of ps for 1 mm gap) and a high counting rate capability. However, a larger amplification of the primary ionization occurs near the cathode resulting in a dependence of the signal on the position of the ionizing event and consequently, in a poor energy resolution (~20%). A hybrid configuration of the detector is MWPC + PPAC or position-sensitive PPAC. The wire structrure of the position-sensitive PPAC gives position information, while it introduces a second step of multiplication near the wires.

Although wire-based detectors have been extensively used in both high-energy and low-energy physics experiments, they present some drawbacks: a) gain limits; b) counting rate limits (space charge); c) aging after long-term operation. The need for gas detectors that could operate at high counting rates led to the development of **micro – strip gas chambers (MSGCs)** [1]. Operation at high counting rate was achieved by increasing the granularity of the anode and by adding cathodes very close to the anodes to evacuate ions during the avalanche process. The electrodes are formed on insulating substrate by micro-lithographic technology. However, MSGCs proved difficult to operate in challenging conditions and were prone to aging and sparking. The invention of the **Micropattern gas detector (MPGD)**, such as the micromesh gaseous structure chamber (the MicroMegas) [2] and the gas-electron multiplier (GEM) detector [3], appears to have solved these problems. MPGDs have small avalanche gaps and therefore a rapid signal development, implemented in slightly different ways. In MicroMegas detectors the electron multiplication takes place in the narrow gap between a thin cathode mesh with holes and the anode. GEMs, on the other hand, have a thin metal-coated polymer (Kapton) foil chemically produced with a high density of holes and the multiplication takes place in the holes in the foil.

The **drift chamber** gives spatial information by measuring the electrons drift time. The time measurement starts by an external (fast) detector, i.e. scintillator counter. The electrons drift to the anode (sense wire) in the field created by the cathodes and their arrival at the anode stops the time measurement. The spatial resolution of the drift chamber depends on: a) the electron diffusion and, thus, on the drift path and b) the time resolution of the electronics, including the start detector.

A combination of MWPC and drift chamber is the **Time Projection chamber** (**TPC**). Its geometry is mostly cylindrical with a central HV cathode, a MWPC at one end-cap of the cylinder and with a magnetic field B parallel to the electric field E. The electrons drift to the end-caps in homogeneous E. The drift distance is of several meters and the amplification of the primary ionization occurs at the end of the drift. With the TPC it is possible to perform a full 3D reconstruction of the particle trajectory, xy coordinates are obtained from the anode wires and cathode pads of MWPC and z is deduced from the drift time measurement. A PID is also possible by measuring the momentum via dE/dx.

For further reading relative to this section, the reader can also refer to [4–9].

2 Gas detectors in nuclear physics experiments with RIBs

In low-energy nuclear physics experiments with RIBs some key features should be taken into consideration for the design of detecting systems: the beams produced by first generation facilities have low intensity (up to 10^7 pps), large energy spread and large emittance (in-flight production) and sometimes a high background. To face these problems the detection system should be highly efficient, with good energy and angle resolution, low energy threshold, able to suppress background (by performing coincidence measurements), to perform event-by-event ion beam tracking and a full kinematic reconstruction. To achieve high statistics, the target should be thick enough, though a thick target results in a poor energy resolution.

Figure 1. (color online) Left panel: Schematic exploded view of the position-sensitive PPACs electrodes. Middle panel: Electrode package. Right panel: Final assembly of the position-sensitive PPAC.

In the following I present the position-sensitive PPACs employed for nuclear physics and nuclear astrophysics experiments with the low-energy RIBs delivered by the in-flight facility EXOTIC (Legnaro National Laboratory of INFN, Italy) [10–13]. The two position-sensitive PPACs are dedicated to the RIB tracking system and to Time of Flight (ToF) measurements. They are placed upstream the reaction target that is installed at the focal plane of the EXOTIC facility. For more details on the experimental set up of the facility see [14]. They have a three-electrode structure (Figure 1, left-hand side): a central cathode and two anodes, placed symmetrically with respect to the cathode at a distance of 2.4 mm. The detector active area is 62 x 62 mm^2. The cathode is made of a 1.5 μm-thick stretched mylar foil with 30 nm of aluminum evaporated on both surfaces. Each anode is a mesh of 60 20 μm-thick wires in the x and y directions, with a spacing of 1 mm. The wires of the first anode are oriented horizontally and those of the second one vertically. The position information of a particle crossing the PPAC is extracted from the anode signals by using a delay-line readout. Each wire is electrically connected to discrete LC circuit delay lines of 2.3 ns/mm each, with a 50 Ω impedance, resulting in a total delay of 138 ns in both the x and the y direction. The cathode signal is used as a reference time for ToF measurements and for trigger purposes. The PPAC is filled with C_4H_{10} (high gain at low pressure) at a working pressure of 10-20 mbar. The cathode is biased at a negative potential of 550-970 V while the anodes are kept at ground potential. In this way, reduced electric fields of 200-230 V/cm/mbar are obtained.

A very important issue for handling the PPAC signals, in order to have a good position resolution and a high tracking efficiency, is the noise reduction that can be reached by performing a careful grounding and by using a low-noise fast preamplifier, placed very close to the PPAC. The cathode signal and the anode signals x_1, x_2, y_1, y_2, extracted from each end of the delay lines are sent to this fast preamplifier, followed by a further amplification in a fast amplifier and then is sent to a Constant Fraction Discriminator (CFD) to set a threshold and extract logic signals (left-hand side of Figure 2). Finally, it is a sent to a Time to Digital Converter (TDC). The position determination of the detected particle is achieved by measuring with a TDC the time interval, t_{x1}, t_{x2}, t_{y1} and t_{y2}, between a common start (given by the master trigger signal) and the delay line outputs. The positions of a particle crossing the PPAC are: $X_{position} = k\ (t_{x1} - t_{x2})/2\ (mm)$ and $Y_{position} = k\ (t_{y1} - t_{y2})/2\ (mm)$, where $k = 0.435$ mm/ns is the slope for the x and y delay lines. The good events retained in the analysis are those lying on the diagonal of the $(t_{x2} - t_K)\ vs\ (t_{x1} - t_K)$ correlation plot (shown on the right-hand side of Figure 2) that satisfy the equation: $total\ delay\ line = (t_{x1} - t_K) + (t_{x2} - t_K)$, with t_K the time interval between the common start and the cathode signal. Selecting these events allows us to reject pile up events and multiple hit events.

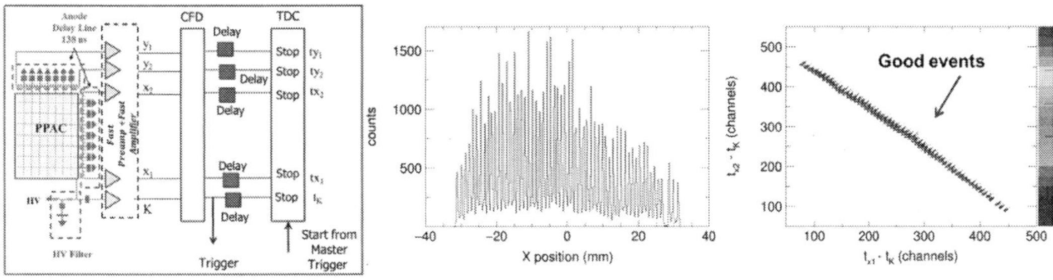

Figure 2. (color online) Left-hand side: Electronics of a position-sensitive PPAC. *X* position (middle) and $(t_{x2} - t_K)\ vs\ (t_{x1} - t_K)$ correlation plot (right-hand side) obtained for an ^{241}Am α source. For details see text.

The PPAC has 0.86 ns FWHM time (intrinsic) resolution, allowing ToF measurements between PPAC and the silicon detectors of the EXPADES array [14] with a FWHM overall time resolution of about 1.5 ns. Furthermore, the 0.7 mm FWHM position resolution of the PPACs (binary readout, FWHM $=2.35\frac{1\ mm}{\sqrt{12}}$) permits reconstruction of the event position on the reaction target with 1.6 mm FWHM resolution. The PPACs were found to be able to sustain high counting rates up to $\sim 10^6$ Hz and to have a tracking efficiency of 90%-98%, depending on the charge (Z=3-9) and the rate of the produced RIB.

3 Active chamber for experiments with low-energy RIBs

In nuclear physics experiments with low-energy RIBs the use of an active target, a device that acts as target and as detector (TPC) simultaneously, presents potentially some interesting features: 4π

geometry; full 3D tracking of all particles (from position resolution is deduced the energy resolution at the vertex); allows to use high thickness of the target without loss of energy resolution (gain up to 100); low energy threshold; possibility to detect also unbound states; very low background in many cases (coincident measurements); excitation function can be done in a single measurement (important for resonant experiments: quasi-molecular structures and experiments of astrophysical interest); versatility.

One of the first generation active targets, MAYA [16], can be seen on the left-hand side of Figure 3. In this device the projected trajectories on the cathode pads (1024 pads) of the charged particles are used to have a 2D image while the third dimension is taken from the electron drift time obtained from the anode wires (32 wires with 2.3-mm gap). The deduced information is angle, energy (from the range or the charge signal) and PID (through the correlation plot range *vs* charge). A wall of ancillary detectors (Si+CsI) at the exit of MAYA are used for the detection of particles that go through. The amplification gain is 10^3-10^4 and the RMS position resolution ~ 1 mm for a pad pitch 0.9 cm. The position resolution determines the angular and range resolution and thus, it determines the energy resolution and the PID for stopped particles. For particles that go through, the PID is obtained via ΔE-E technique.

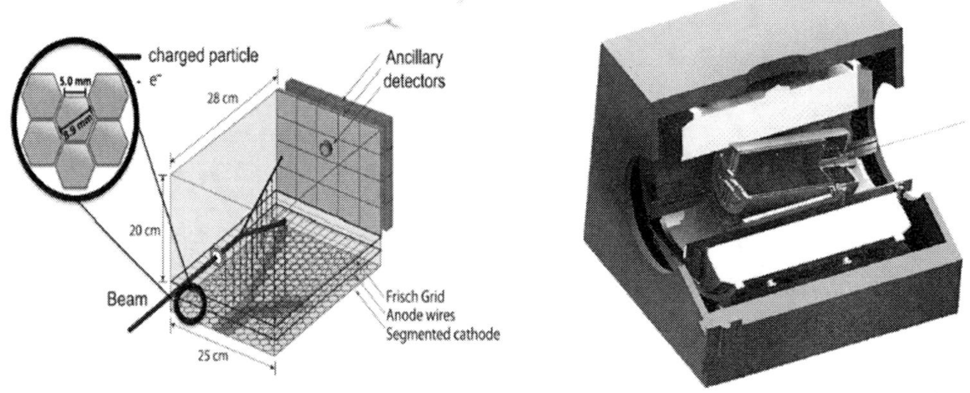

Figure 3. (color online) Left-hand side: MAYA [16]. Right-hand side: AT-TPC [17].

There are some limits in such a first generation active target: poor dynamic range ~ 20; can sustain a limited rate due to the electronics high dead time > 2 ms; position resolution since it is hard to reconstruct trajectories for ranges of few cm. Finally, only two-body reactions can be considered because the drift time deduced from wires does not allow dealing with multi-hit events.

For more intense RIBs (provided by facilities like SPES, SPIRAL2, ISOLDE etc), second generation active targets are required to overcome the above described limitations. Switching from wire-based amplification to amplification based on MPGDs renders second generation devices robust, eliminates ion backflow permitting large range of beam rates, allows high rate and high dynamic range, dealing with multi-hit events, better position resolution, better PID, larger target thickness (from 25 to 100 cm), more versatility. The trigger can be both internal and external. The use of GET (General electronics per TPCs) allows to handle up to 30000 channels.

Various second generation active targets have been already built or are under construction (for a detailed list see [17]). Here, I just refer to ACTAR (in construction, funded by ERC Starting Grant, G.F. Grinyer) and AT-TPC in construction at NSCL (FRIB, USA) [17]. ACTAR is cubic or cuboid with about 16000 channels and amplification of MicroMegas-type. It uses GET electronics and auxiliary particle detectors are employed for the detection of charged particles that go through (Si+Si or Si+CsI) and for γ rays ($LaBr_3$ or $CeBr_3$). AT-TPC (see right-hand side of Figure 3) is a cylinder (l=1 m, ϕ=55 cm for an active volume of 250 l) inside a 2T solenoid. This allows the measurement of the particle momentum from the curvature giving a better energy resolution compared with that deduced from the particle range and, moreover, it allows the characterization of energetic particles leaving the volume. There is no need for ancillary detectors and the trigger is internal. The employed magnetic field increases the range to stop the particles and focuses drift electrons yielding in a better localization. The amplification is also in this case of MicroMegas type. The axis of the detector is tilted by 7° to spread tracks for small angles permitting pile up reduction by a factor about 500.

4 Summary

I presented the operation principle and different gas detecting systems, focusing on the requirements of nuclear physics experiments with low-energy RIBs. An example of gas detector, operating at the in-flight RIB facility EXOTIC was described: position-sensitive PPAC for the RIB tracking and for ToF measurements. Moreover, an interesting device for nuclear physics experiments with RIBs was discussed: the active target. Both first and second generation active targets were presented, describing the achieved improvements for addressing challenges with more intense RIBs, delivered from second generation facilities.

References

[1] A. Oed, NIMA **263** (1988) 351
[2] Y. Giomataris et al. NIM A **376** (1996) 29
[3] F. Sauli, NIMA **386** (1997) 531
[4] "Techniques for nuclear and particle physics experiments", W. R. Leo
[5] "Radiation Detection And Measurement", G. Knoll
[6] "Tracking Detectors", Lecture, N. Wermes, SLAC Summer Institute 2016 "New Orizons on the energy frontier"
[7] "Gaseous detectors for radioactive beam experiments", D. Suzuki and "Trend and new developments in gas amplifiers: a concise review", M. Cortesi, GDS Topical Meeting: GDS coupling to auxiliary detection systems, January 25-27, 2017 INFN Laboratori Nazionali di Legnaro
[8] "Making radioactive ion beams - Detecting reaction products ", R. Raabe, Rewriting Nuclear Physics Textbooks 30 years with Radioactive Ion Beams Physics, Pisa, July 2015
[9] F. Sauli, Physics Reports 403–404 (2004) 471–504
[10] V.Z. Maidikov et al., Nucl. Phys. A **746** (2004) 389c
[11] D. Pierroutsakou et al., Eur. Phys. J. Special Topics **150** (2007) 47
[12] F. Farinon et al., Nucl. Instr. and Meth. B **266** (2008) 4097
[13] M. Mazzocco et al., Nucl. Instr. and Meth. B **266** (2008) 4665; Nucl. Instr. and Meth. B **317** (2013) 223
[14] D. Pierroutsakou et al., Nucl. Instr. and Meth. A **834** (2016) 46
[15] D. Guillemaud-Mueller et al., IEEE Trans. Nucl. Sci. **33** (1) (1986) 343
[16] C.E. Demonchy et al., Nucl. Instr. Meth. A **583** (2007) 341
[17] S. Beceiro-Novo, T. Ahn, D. Bazin, W. Mittig, Progr. Part. and Nucl. Phys. **84** (2015) 124–165

The Trojan Horse Method in Nuclear Astrophysics

Aurora Tumino[1,2,⋆], *Claudio* Spitaleri[2], *Silvio* Cherubini[2,3], *Giuseppe* D'Agata[2], *Giovanni Luca* Guardo[2], *Marisa* Gulino[1,2], *Iolanda* Indelicato[2], *Marco* La Cognata[2], *Livio* Lamia[2], *Rosario Gianluca* Pizzone[2], *Giuseppe Gabriele* Rapisarda[2], *Stefano* Romano[2,3], *Maria Letizia* Sergi[2], and *Roberta* Spartá[2]

[1] *Facoltá di Ingegneria e Architettura, Universitá degli Studi di Enna "Kore", Enna, Italy*
[2] *INFN-Laboratori Nazionali del Sud, Catania, Italy*
[3] *Dipartimento di Fisica e Astronomia, Universitá degli Studi di Catania, Catania, Italy*

Abstract. The Trojan Horse Method (THM) represents the indirect way to measure reactions between charged particles at astrophysical energies. This is done by measuring the quasi free cross section of a suitable three body process. The basic features of the THM will be presented together with some applications to demonstrate its practical use.

1 Introduction

The Trojan Horse Method (THM) has been introduced in nuclear astrophysics as an indirect approach to determine low-energy cross sections overcoming the main issues of direct experiments, such as the Coulomb repulsion and the electron screening [1, 2]. The first one is responsible for the exponential decrease of the cross section at the relevant temperatures, while the electron screening, due to the electrons surrounding the interacting ions, leads to an increased cross section compared to the one for bare nuclei that is necessary to assess the reaction rate in astrophysical plasma.
The THM ([3–6] and references therein) makes use of a suitable $A + a \to b + B + s$ two-to-three body process to measure the astrophysical $A + x \to b + B$ two-body reaction of interest by means of a relation between the two based on nuclear reaction theories. The choice of the three-body process is done in such a way that target a (or equivalently the projectile) has a wave function with a large amplitude for a $x - s$ cluster configuration, x being the target (or equivalently the projectile) of the two-body reaction. The selected part of the three-body phase space conveys with the quasi-free (QF) kinematics: the other cluster s remains spectator to the process, and $A + x \to b + B$ can be regarded as a half-off-energy-shell (HOES) two-body reaction, usually referred to as a QF reaction. The QF reaction mechanism is sketched in Fig.1 using a pole diagram with upper and lower vertexes describing the break-up of a and the $A + x \to b + B$ two-body reaction, respectively.

Since the three-body process occurs at an energy above the Coulomb barrier, the main feature is the real suppression of both Coulomb barrier and screening effects in the HOES two-body cross section. Nevertheless, the quasi-free $A + x$ process can occur even at very low sub-Coulomb energies because the $A + a$ relative motion is compensated for by the $x - s$ binding energy. This is indeed a different approach to the THM [3] compared to the original idea of Baur [7], where the initial velocity

⋆e-mail: tumino@lns.infn.it

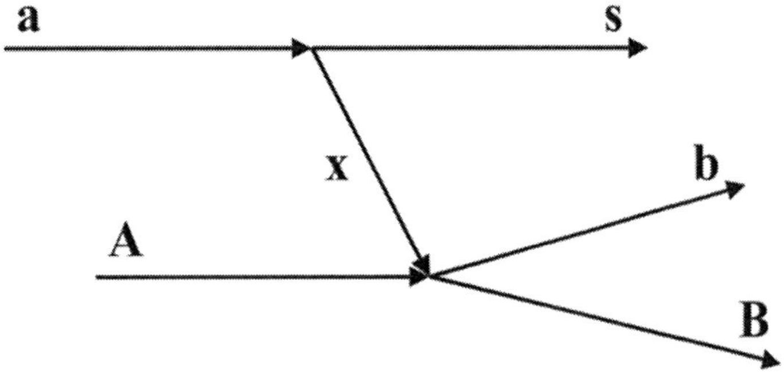

Figure 1. Pole diagram describing the QF mechanism in the $A + a \to b + B + s$ reaction.

of the projectile A is compensated for by the $x - s$ intercluster motion. In that framework a quite large momentum of the order of 200 MeV/c or more is needed. But the relative yield of the experimental momentum distribution at such momenta can be very small, in particular for a l=0 inter-cluster motion (for example p-n motion inside ^{2}H or α-d motion inside ^{6}Li). This would complicate the separation from other competitive reaction mechanisms. Moreover, the theoretical description of the tails of the momentum distribution is a hard task, their shape being very sensitive to it. In our approach to the THM, the intercluster motion is only needed to determine the accessible astrophysical energy region. It corresponds to a cutoff in momentum distribution of s of few tens of MeV/c. In this framework, the so called "quasi-free two-body energy" is given by:

$$E_{QF} = \frac{m_x}{m_x + m_A} E_A - B_{x-s}. \tag{1}$$

where E_A represents the beam energy, m_x and m_A are the masses of x and A particles respectively, and B_{x-s} is the binding energy for the $x - s$ system.
In the Impulse Approximation, based essentially on the assumption that the interaction of the spectator with particles b and B is neglected, the three body-cross cross section can be factorized as:

$$\frac{d^3\sigma}{dE_b d\Omega_b d\Omega_B} \propto [KF\ |\varphi_a(\mathbf{p}_{sx})|^2] \left(\frac{d\sigma}{d\Omega_{c.m.}}\right)^{\mathrm{HOES}} \tag{2}$$

where KF is a kinematical factor containing the final state phase-space factor. It is a function of the masses, momenta and angles of the outgoing particles [25]; $\varphi_a(\mathbf{p}_{sx})$ is proportional to the Fourier transform of the radial wave function $\chi(\mathbf{r})$ for the $x - s$ inter-cluster relative motion; $(d\sigma/d\Omega_{c.m.})^{\mathrm{HOES}}$ is the half-off-energy-shell (HOES) differential cross section for the binary reaction at the center of mass energy $E_{c.m.}$ given in post-collision prescription by

$$E_{c.m.} = E_{bB} - Q_{2b}\ . \tag{3}$$

Here, Q_{2b} is the Q-value of the binary reaction and E_{bB} is the relative energy of the outgoing particles c and C, which spans the accessible astrophysical region defined above.

The factorization of Eq.2 is strictly valid in Plane Wave Impulse Approximation, which changes only the absolute magnitude but not the energy dependence of the two-body cross section.

2 How to apply the THM

In THM experiments, any two of the three particles in the exit channel are usually detected and identified by means of telescopes (silicon detector or ionization chamber as ΔE detector and PSD as E detector) placed in a phase space region where quasi free kinematics dominates. After the reconstruction of the reaction channel, a critical point is the selection of the quasi free mechanism from other reaction mechanisms feeding the same particles in the final state, e.g. sequential decay and direct break-up. This is done investigating the shape of the experimental momentum distribution $\Phi(p_s)$ of the spectator by means of the energy sharing method [8], for example. The quasi-free coincidence yield with a cutoff in relative energy of few tens of keV and in center of mass angle of few degrees at most is divided by the kinematic factor, providing a quantity which is proportional to the product of the spectator momentum distribution with the differential two-body cross section (see Eq.2). The window is chosen in such a way that the differential two-body cross section in this range can be considered almost constant. Thus, the quantity defined above represents essentially the momentum distribution for the spectator that in PWIA can be compared with the Fourier transform of the radial $x - s$ bound state wave function.

Data analysis is limited to the region where the agreement between the two distributions is found. Usually a window of few tens of MeV/c is chosen not to move too far from E_{QF}, according to the prescriptions reported in [15].

Therefore, $(d\sigma/d\Omega_{cm})^{\mathrm{HOES}}$ can be extracted from the three-body coincidence yield by simply inverting Eq.2. The Coulomb barrier in $(d\sigma/d\Omega_{cm})^{\mathrm{HOES}}$ is suppressed [9] and this is due to the virtuality of particle x. This seems to be the only consequence of off-energy-shell effects as suggested by the agreement between HOES and OES cross-sections for the ^{6}Li(n,α)^{3}H reaction [10].

Thus, in a final step, to relate the HOES excitation function to the relevant on-energy-shell (OES) one, Coulomb corrections have to be considered. If one looks at the angular distributions no correction is needed because once the energy is fixed, it would mean to introduce simply a scaling factor. Thus, the OES data are directly comparable with HOES ones projected onto the emission angle of C (or c) in the $C - c$ center of mass system, $\theta_{c.m.}$, as given by [14]:

$$\theta_{c.m.} = \arccos \frac{(\mathbf{v}_A - \mathbf{v}_x) \cdot (\mathbf{v}_c - \mathbf{v}_C)}{|\mathbf{v}_A - \mathbf{v}_x||\mathbf{v}_c - \mathbf{v}_C|} \tag{4}$$

where the vectors $\mathbf{v}_A$, $\mathbf{v}_x$, $\mathbf{v}_c$, $\mathbf{v}_C$ are the velocities of projectile, transferred particle and outgoing nuclei respectively. These quantities are calculated from their corresponding momenta in the lab-system, where the momentum of the transferred particle is equal and opposite to that of s when the quasi-free break-up takes place in the target, otherwise a little different formula has to be used [14]. If HOES data are projected onto the E_{Cc} axis, Coulomb suppression has to be introduced before comparison with OES data [11, 12]. In a simple approach, this is done by means of the penetrability factor:

$$P_l(k_{Ax}R) = \frac{1}{G_l^2(k_{Ax}R) + F_l^2(k_{Ax}R)} \tag{5}$$

with F_l and G_l regular and irregular Coulomb wave functions and R the so called cutoff radius, which is usually chosen as the sum of the radii of nuclei A and x. This procedure does not allow us to extract the absolute value of the two-body cross section. However, this is not a real problem since the absolute magnitude can be derived from a scaling to the direct data available at higher energies.

If a resonant two-body reaction is involved, a dedicated approach called Modified R-matrix ([16] and references therein) has been developed to extract the resonance parameters from the THM reaction yield. The considerable advantage is that they are the same needed to calculated the OES cross section.

Table 1. Two-body reactions studied via the THM with measured two-to-three TH reaction and relevant references for each reaction

Direct reaction	TH reaction	ref
^{7}Li(p,α)^{4}He	^{7}Li(d,α α)n	[17–19]
^{7}Li(p,α)^{4}He	^{7}Li(^{3}He,α α)^{2}H	[20]
^{6}Li(d,α)^{4}He	^{6}Li(^{6}Li,α α)^{4}He	[21]
^{6}Li(p,α)^{3}He	^{6}Li(d,α ^{3}He)n	[22–24]
^{11}B(p,α)^{8}Be	^{11}B(d ,^{8}Be α)n	[25, 26]
^{10}B(p,α)^{7}Be	^{10}B(d,^{7}Be α)n	[27–29]
^{9}Be(p,α)^{6}Li	^{9}Be(d,^{6}Li α)n	[30, 31]
^{2}H(^{3}He,p)^{4}He	^{6}Li(^{3}He,p α)^{4}He	[32]
^{2}H(d,p)^{3}H	^{2}H(^{6}Li,t p)^{4}He	[33–35]
^{15}N(p,α)^{12}C	^{15}N(p,α^{12}C)n	[36]
^{18}O(p,α)^{15}N	^{18}O(p,α^{15}N)n	[37, 38]
^{1}H(p,p)^{1}H	^{2}H(p,pp)n	[11, 12]
^{2}H(d,p)^{3}H	^{2}H(^{3}He,t p)^{1}H	[39–41]
^{2}H(d,n)^{3}He	^{2}H(^{3}He,^{3}He n)^{1}H	[39–41]
^{19}F(p,α)^{16}O	^{2}H(^{19}F,α^{16}O)n	[42, 43]
^{17}O(p,α)^{14}N	^{2}H(^{17}O,α^{14}N)n	[44]
^{4}He(^{12}C,^{12}C)^{4}He	^{6}Li(^{12}C,α^{12}C)^{2}H	[45]
n(^{6}Li,t)^{4}He	^{2}H(^{6}Li,t^{4}He)^{1}H	[10, 46]
^{13}C(α,n)^{16}O	^{6}Li(^{13}C,n^{16}O)^{2}H	[16]
^{18}F(p,α)^{15}O	^{2}H(^{18}F,α^{15}O)n	[47]
^{19}F(α,p)^{22}Ne	^{6}Li(^{19}F,p^{22}Ne)^{2}H	[48]
^{7}Be(n,α)^{4}He	^{2}H(^{7}Be,$\alpha\alpha$)^{1}H	[49]
^{12}C(^{12}C,α)^{20}Ne	^{12}C(^{14}N,α^{20}Ne)^{2}H	[50]
^{12}C(^{12}C,p)^{23}Na	^{12}C(^{14}N,p^{23}Na)^{2}H	[50]

3 Resent results

The THM has applied already many times to resonant and non resonant charged particle reactions connected with fundamental astrophysical problems, from primordial to stellar nucleosynthesis. A list of reactions studied by means of the THM is given in Table 1 together with the relevant references.

Given the well assessed THM experimental and theorerical methodology to investigate low-energy nuclear reactions, applications in the near future are forseen for neutron induced reactions on radioactive nuclei. Having a radioactive beam, deuteron targets will serve as a virtual source of neutrons. This turns out to be a unique way to investigate such reactions. Moreover, after the very successful work to determine the C-burning reaction rate, accepted for publication in Nature [50], the THM will be further applied to other key processes triggering the evolution of massive stars, such as $^{12}C+^{16}O$ and $^{16}O+^{16}O$.

References

[1] H.J. Assenbaum et al., *Z. Phys.* A **327** 461 (1987)
[2] F. Strieder et al., *Naturwissenschaften* **88** 461 (2001)
[3] C. Spitaleri et al., Phys. At. Nucl., **74**, 1763, (2011)
[4] A. Tumino et al., *Few Body Syst.*, **54**, Issue 5-6 745 (2013)
[5] A. Tumino et al., *Few Body Syst.*, **54**, Issue 7-10 869 (2013)
[6] Tribble R. et al., *Rep. Prog. Phys.* **77**, Issue: 10 106901 (2014)
[7] G. Baur, *Phys. Lett.* B **178** 135 (1986)
[8] J. Kasagi et al., *Nucl. Phys.* A **239** 233 (1975)
[9] A.M. Mukhamedzhanov et al., *Eur. Phys. J.* A **27** 205 (2006)
[10] A. Tumino et al., *Eur. Phys. J.* A **25** 649 (2005)
[11] A. Tumino et al., *Phys. Rev. Lett.* **98** 252502 (2007)
[12] A. Tumino et al., *Phys. Rev.* C **78** 064001 (2008)
[13] M. La Cognata et al., *Astrophysical J.* **777**, 143 (2013)
[14] M. Jain et al., *Nucl. Phys.* A **153** 49 (1970)
[15] Shapiro, I.S. et al., *Nucl. Phys.* A, **61**, 353 (1965)
[16] M. La Cognata et al., *Phys. Rev. Lett.* **49** 015106 (2013)
[17] C. Spitaleri et al., *Phys. Rev.* C **60** 055802 (1999)
[18] M. Aliotta et al. 2000 *Eur. Phys. J.* A **9**, 435
[19] M. Lattuada et al., Ap.J. **562**, 1076 (2001)
[20] A.Tumino et al., *Eur. Phys. J.* A direct, 1 (2006) DOI: 10.1140/epja/i2006-08-038-1
[21] C. Spitaleri et al., *Phys. Rev.* C **63** 005801 (2001)
[22] A. Tumino et al., *Phys. Rev.* C **67** 065803 (2003)
[23] A. Tumino et al., *Prog. Theor. Phys. Suppl.* **154** 341 (2004)
[24] L. Lamia et al., *Astrophysical J.* **768** 65 (2013)
[25] C. Spitaleri et al., *Phys. Rev.* C **63** 055806 (2004)
[26] L. Lamia et al., *Journ. of Phys. G: Nucl. Part. Phys.* **39** 015106 (2012)
[27] L. Lamia et al., *Nucl. Phys. A* **787** 309 (2007)
[28] L. Lamia et al., *Il Nuovo Cimento* **31** N.4 423 (2008)
[29] C. Spitaleri *et al.*, *Phys. Rev.* C **95** 035801 (2017)
[30] Qun-Gang Wen et al. *Phys. Rev.* C **78** 035805 (2008)
[31] Q. Wen *et al. J.Phys. G: Nucl. Part. Phys.*, **38**, 085103 (2011)
[32] M. La Cognata et al., *Eur. Phys. J.* A **27** 249 (2006)
[33] A. Rinollo et al. *Nucl. Phys.* A **758** 146 (2005)
[34] Li Chengbo *et al. Phys. Rev.* C **92** 025805 (2015)
[35] Li Chengbo *et al. Phys. Rev.* C **95** 035804 (2017)
[36] M. La Cognata et al., *Phys. Rev.* C **76** 065804 (2007)
[37] M. La Cognata et al., *Phys. Rev. Lett.* **101** 152501 (2008)
[38] M. La Cognata *et al.*, *Astrophysical J.* **708** 796 (2010)
[39] A. Tumino *et al. Phys. Lett.* B **700** 111 (2011)

[40] A. Tumino *et al. Phys. Lett.* B **705** 546 (2011)
[41] A. Tumino *et al., Astrophysical J.* **785** 96 (2014)
[42] M. La Cognata *et al., Astrophysical J. Lett.* **739** L54 (2011)
[43] I. Indelicato *et al., Astrophysical J.* **845** 19 (2017
[44] M.L. Sergi *et al., Phys.Rev.* C **R82** 032801 (2010)
[45] C. Spitaleri *et al., Eur. Phys. Journ.* A**7** 181 (2000)
[46] M. Gulino *et al., J.Phys. G: Nucl. Part. Phys.*, **37** 125105 (2010)
[47] S. Cherubini *et al., Phys.Rev.* C **R92** 015805 (2015)
[48] R.G. Pizzone *et al., Astrophysical J.* **836** 57 (2017)
[49] L. Lamia *et al., Astrophysical J.* **850** 175 (2018)
[50] A. Tumino *et al., Nature* 2018 in press

Direct measurements and detection techniques with low-energy RIBs

H. Yamaguchi[1,⋆], *S.* Hayakawa[1], *L.* Yang[1], *H.* Shimizu[1], *D.* Kahl[2,1], and *CRIB collaboration*

[1] *Center for Nuclear Study (CNS), University of Tokyo, RIKEN campus, 2-1 Hirosawa, Wako, Saitama 351-0198, Japan*
[2] *School of Physics and Astronomy, the University of Edinburgh, Peter Guthrie Tait Road, Edinburgh EH9 3BF, UK*

Abstract. Astrophysical reactions involving radioactive isotopes (RI) are of importance for the stellar energy generation and nucleosynthesis especially in high-temperature astrophysical sites, such as X-ray bursts, core-collapse supernovae, and supermassive metal-poor stars. In spite of the essential difficulties in the experimental evaluation of those reaction rates, there are several successful approaches to study them, owing to the recent technical developments in the beam production, measurement method, and detectors. Among them, the measurements of α resonant scattering and (α, p) reactions using the thick-target method in inverse kinematics are discussed. The experiments at the low-energy RI beam separator CRIB, operated by Center for Nuclear Study (CNS), the University of Tokyo, are introduced as examples for such studies.

1 Introduction

The essential difficulty in experimental nuclear astrophysics lies in reproducing the special conditions of stellar environments. In stars, vast amount of atoms can be condensed with an extremely high density and nuclear reactions with tiniest cross sections can be effective. Some processes are slow and may take more than millions of years, which is much longer than our life, but stars can live such long to complete the process. The typical temperature of stellar burning sites is 10^7–10^9 K, which can be regarded as a high temperature, but lower than the temperature corresponding to the Coulomb barrier of the nucleus. This implies the astrophysical reactions may occur slowly, making the experiment to simulate them extremely difficult.

Some astrophysical reactions involve radioactive isotopes (RI). Although the RI are seldom seen on the earth, they may play an important role in stars. An example is the CNO cycle, which is a basic energy generation process in stars, taking place even in our sun. In the CNO cycle, the radioactive ^{13}N nucleus is produced via the ^{12}C$(p, \gamma)^{13}$N reaction. The β-decay half life of ^{13}N is 10 minutes, which may restrict the speed of the cycle. If the cycle includes ^{14}O by the ^{13}N$(p, \gamma)^{14}$O reaction (the hot-CNO cycle), the slow β-decay of ^{13}N is bypassed, and the cycle is accelerated, resulting in a higher energy generation rate.

⋆e-mail: yamag@cns.s.u-tokyo.ac.jp

The earliest study of astrophysical reaction with RI beam was in fact on this $^{13}N(p, \gamma)^{14}O$ reaction [1, 2]. Such measurements with RI beam often suffer from the limitation of beam intensities. The intensity of RI beam is typically 10^5 particles per second (pps), while $>10^{14}$ pps is available for light-ion beams. This great difference in the beam intensity is fundamental for the feasibility of the measurement. Another limitation is the target. In a normal experimental condition, short-lived RI or neutrons can not be used as the target, and can only be used as the beam.

As far as we discuss astrophysical implication of RI-involving reactions, there are two favorable facts which should be noted: 1) In most cases the RI in stars play an effective role in high-temperature (explosive) phenomena, where the reaction cross sections are not too small. 2) We may think about some reaction involving very exotic nuclei and having a very small cross section, which is extremely difficult to be measured. In such a case, however, the reaction should be of small or no importance for the astrophysics.

In spite of the technical limitation, there are several successful experimental approaches:

- Direct measurement for studying high-temperature phenomena
  This is feasible with relatively high-intensity RI beam (^{7}Be, ^{13}N [2] etc.), or high-cross section reactions (such as (α, p) reactions).
- Studying resonance properties
  Some reactions proceed predominantly via resonant reactions, and we only need to study the property of resonances by transfer reactions [3], resonant scatterings [4], or some other methods.
- Indirect measurement with RI beams
  Indirect measurements with the Asymptotic Normalization Coefficient (ANC) [5, 6], Coulomb breakup [1, 7] or Trojan horse method [8] can be carried out even with RI beams.

In this article, we particularly discuss on the measurement of ^{7}Be+α resonant scattering and the direct measurement of $^{11}C(\alpha, p)$ reaction with the thick-target method in inverse kinematics (TTIK) as successful examples.

2 Thick-target method in inverse kinematics

The TTIK [9] is a method suitable for astrophysical reaction studies. In that method, the beam energy is degraded in a thick reaction target, and the reaction occurs at various center-of-mass energies. By detecting the position and the energy of light particles emitted after the reaction, the reaction position in the target and the reaction energy are deduced according to the kinematical relationship. This method has several advantages, namely, (a) using inverse kinematics, we can study reactions with short-lived RI which cannot be used as the target, (b) we can perform a simultaneous measurement of cross section for a certain energy range without varying the incoming RI beam energy, (c) when the beam is stopped in the target, we can perform measurements at 180° in center-of-mass angle, where the Coulomb scattering is minimal, and (d) the inverse kinematics condition can be helpful for a high resolution measurement at the energy region lower than the beam energy.

Many proton- and α-resonant scattering [4, 10–13] experiments have been performed with the TTIK method at CRIB [14, 15], which is an RI beam separator operated by Center for Nuclear Study (CNS), the University of Tokyo. Most of those studies are related to the astrophysical (p, γ) or (α, γ) capture reactions. Direct measurement of (α, p) reactions, such as $^{14}O(\alpha, p)$ [16], $^{11}C(\alpha, p)$ [17], $^{21}Na(\alpha, p)$, $^{18}Ne(\alpha, p)$, and $^{22}Mg(\alpha, p)$, have also been performed with the TTIK method using RI beams at CRIB.

2.1 ^{7}Be+α elastic resonant scattering

A typical example of the resonant scattering experiment at CRIB is for the ^{7}Be+α system. The measurement allows us to evaluate the rate of the ^{7}Be(α, γ) reaction, which is considered to play an important role in the hot *p-p* chain and related reaction sequences [18]. Several reaction sequences including the ^{7}Be(α, γ) reaction should take place in some high-temperature environments at $T_9 > 0.2$, where T_9 is the temperature in GK. During the νp-process in core-collapse supernovae [19], the ^{7}Be(α, γ) reaction may contribute as much as the triple-α process to the synthesis of elements heavier than boron at the relevant temperature of T_9 = 1.5–3 [20]. The Gamow energy window at the highest temperature T_9 = 3 corresponds to the excitation energy E_{ex}=8.2–9.6 MeV in ^{11}C.

In the measurement [4], a low-energy ^{7}Be beam at 14.7 MeV was produced using a 2.3-mg/cm^2-thick hydrogen gas target and a primary ^{7}Li beam at 5.0 MeV/u. The typical ^{7}Be beam intensity was 1–2 $\times$ 10^5 per second at the secondary target, and the main measurement using a thick helium-gas target was performed for 4 days, injecting 2.9 $\times 10^{10}$ ^{7}Be particles into the target. We obtained an excitation function of the elastic scattering with several peaks corresponding to the resonance structure in ^{11}C. The obtained excitation function is shown in the left panel of Figure 1. An R-matrix analysis was performed to deduce the parameters of the resonances, as the calculated curve also shown in the figure. A similar measurement was independently carried out by M. Freer *et al.* at other facilities [21], but our measurement included γ-ray detection to identify inelastic scattering events, and several differences were found in the obtained spectra [4], such as a shift in resonance energy by 500 keV and almost factor 2 difference in the absolute cross section.

The observed resonances may contribute to the astrophysical ^{7}Be(α, γ)^{11}C reaction rate at high temperature, $T_9 > 1.5$. We calculated the resonant reaction rate and compared it with the evaluation in NACRE [22, 23]. In the NACRE evaluation, only 2 resonances at 8.1045 and 8.420 MeV are included. These two resonances dominate the reaction rate $N_{\mathrm{A}}\langle\sigma v\rangle$ up to the temperature $T_9 \sim 3$, and a Hauser-Feshbach calculation was used to provide the reaction rate at higher temperatures. The resonant reaction rates were calculated for four resonance parameter sets using analytical formula described in [22], as in Fig. 1. In conclusion, the resonances at 8.90 MeV and 9.20 MeV have a possibility to give significant contributions to the reaction rate for T_9 = 1.5–3, although they are unlikely to be comparable to the dominant contribution of the 8.420-MeV resonance.

2.2 Direct measurement of the ^{11}C(α, p) reaction

(α, p) reactions often play an important role in the burning process [24] taking place at high temperature stellar sites, such as, X-ray bursts, core-collapse supernovae, and supermassive metal-poor stars. ^{11}C(α, p) is one of the key reactions in the νp-process [20], and the first direct measurement was performed at CRIB [17] with the TTIK. The experimental setup was essentially similar to the α resonant scattering, as shown in Figure 2.

The TTIK is an efficient method as it allows us to measure reaction events occurring at various energies simultaneously. However, it implies the identification of the exit channel may not be complete, because the particles with different origins (such as elastic scattering and inelastic scattering) can be detected at the same detector position with the same energy. This is not a crucial problem for the resonant elastic scattering, in which the cross section is much enhanced by the interference between the nuclear and Coulomb scatterings. For the (α, p) reactions, however, the reaction with the transition to the excited states in the residual nucleus should be accounted for.

With the setup as shown in Figure 2 we identified the reactions with the transition to different states in the excited ^{14}O, namely, ^{11}C(α, p_0), ^{11}C(α, p_1), and ^{11}C(α, p_2) with the time-of-flight information. We evaluated the contribution of each of these reactions separately, and concluded that the (α, p_1) and

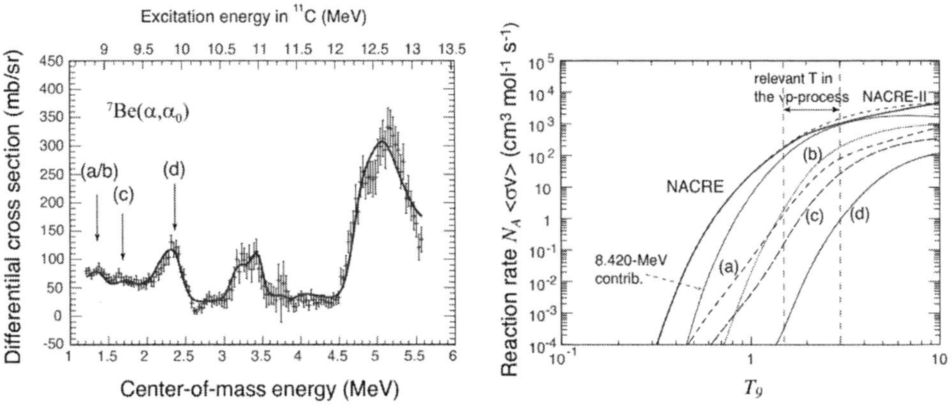

Figure 1. Excitation function of ^{7}Be+α elastic scattering with an R-matrix fit curve (left panel) and evaluated resonant reaction rates of the ^{7}Be(α, γ) reaction for the 8.90, 9.20, and 9.97-MeV resonances, calculated by the analytical formula. (a) and (b) are for the same resonance, but with different spin-parity. The evaluation by NACRE [22] and NACRE-II [23] are shown for comparison. The contribution by the 8.420-MeV resonance, included in NACRE, is also shown.

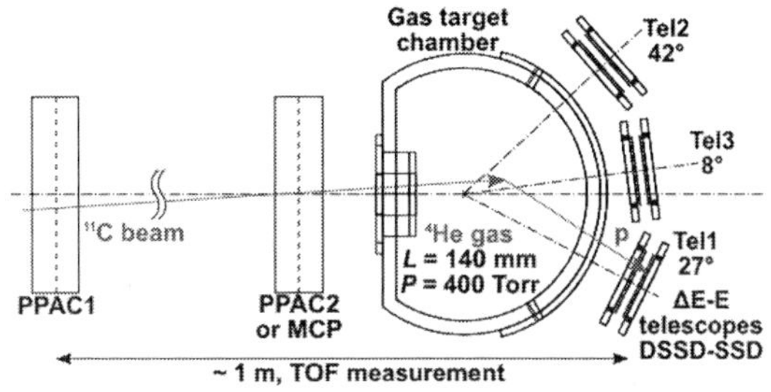

Figure 2. Experimental setup for the ^{11}C(α, p) reaction measurement [17]

(α, p_2) reactions may enhance the reaction rate by 20% at most (see [17] for details). The reaction had been studied also by the reverse ^{14}N(p_0, α) reaction [25], however, such contribution could only be evaluated by the ^{11}C(α, p) reaction measurement.

3 Active target for an advanced form of TTIK

We have developed an active target, referred to as "GEM-MSTPC" [26], for an improved TTIK measurement. That active target is basically a time projection chamber (TPC) with gas electron multiplier (GEM) readout, operated with helium (90%) + CO_2 (10%) gases in the chamber. The helium gas serves as the reaction target and the detector gas simultaneously, and we can detect the 3-dimensional trajectories of the particles directly in the gas target. For the TTIK method, this active target provides additional information compared to the static-target experiment; the trajectory and energy loss for each of the beam particle and the reaction products. This covers the disadvantage of the normal thick target method, the potential misidentification of reactions when there are two or more possible reactions.

Several measurements have been performed with this active target, including the ^{30}S+α elastic scattering [13]. In the ^{30}S+α experiment, we successfully obtained the Bragg curve of the beam particle in the target, and the elastic scattering events were clearly identified. However, the advantage of the active target was preliminary in that experiment, because of the limited energy and position resolution of TPC. Nevertheless, new resonances are observed and we evaluated the resonant contribution to the ^{30}S(α, p) reaction, which is one of the most effective reaction for the energy generation in X-ray bursts [27].

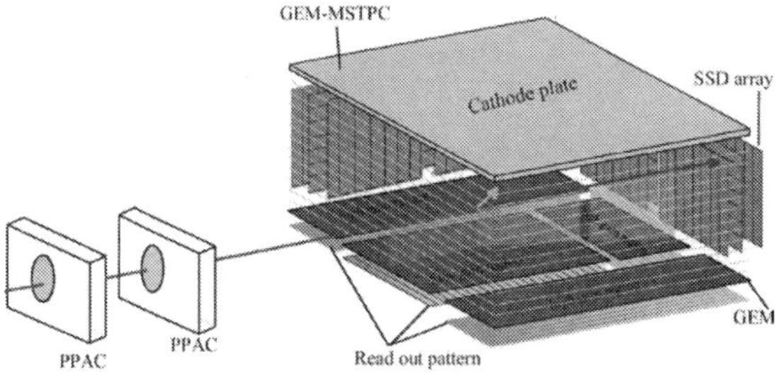

Figure 3. Active target "GEM-MSTPC" of CRIB

4 Summary

In spite of the technical difficulties in studying astrophysical reactions involving RI, there are several approaches for the determination of reaction rates directly or indirectly. The TTIK is one striking method to study astrophysical reactions with RI beams. So far, proton or α resonant scattering experiments to study properties of resonances, and direct measurement of (α, p) reactions are successfully applied for the experimental evaluation of astrophysical reaction rates. More advanced techniques for the RI beam production, measurement scheme and detection system would be required for a systematic understanding of RI-involving astrophysical reactions. The active target is one such technique which can cover the present disadvantages of the TTIK.

Acknowledgments

The experiments were performed at RI Beam Factory operated by RIKEN Nishina Center and CNS, the University of Tokyo. We are grateful to the RIKEN and CNS accelerator staff for their help. This work was partly supported by JSPS KAKENHI (Nos. 25800125, 15K17662, 16K05369, and 16H03980) from the Ministry of Education, Culture, Sports, Science and Technology (MEXT) of Japan.

References

[1] T. Motobayashi, T. Takei, S. Kox, C. Perrin, F. Merchez, D. Rebreyend, K. Ieki, H. Murakami, Y. Ando, N. Iwasa et al., Physics Letters B **264**, 259 (1991)

[2] P. Decrock, T. Delbar, P. Duhamel, W. Galster, M. Huyse, P. Leleux, I. Licot, E. Liénard, P. Lipnik, M. Loiselet et al., Phys. Rev. Lett. **67**, 808 (1991)

[3] D.W. Bardayan, J.A. Howard, J.C. Blackmon, C.R. Brune, K.Y. Chae, W.R. Hix, M.S. Johnson, K.L. Jones, R.L. Kozub, J.F. Liang et al., Phys. Rev. C **74**, 045804 (2006)
[4] H. Yamaguchi, D. Kahl, Y. Wakabayashi, S. Kubono, T. Hashimoto, S. Hayakawa, T. Kawabata, N. Iwasa, T. Teranishi, Y. Kwon et al., Phys. Rev. C **87**, 034303 (2013)
[5] Z.H. Li, B. Guo, S.Q. Yan, G. Lian, X.X. Bai, Y.B. Wang, S. Zeng, J. Su, B.X. Wang, W.P. Liu et al., Phys. Rev. C **74**, 035801 (2006)
[6] O.R. Tojiboev, R. Yarmukhamedov, S.V. Artemov, S.B. Sakuta, Phys. Rev. C **94**, 054616 (2016)
[7] F. Schümann, S. Typel, F. Hammache, K. Sümmerer, F. Uhlig, I. Böttcher, D. Cortina, A. Förster, M. Gai, H. Geissel *et al.*, Phys. Rev. C **73**, 015806 (2006)
[8] S. Cherubini, M. Gulino, C. Spitaleri, G.G. Rapisarda, M. La Cognata, L. Lamia, R.G. Pizzone, S. Romano, S. Kubono, H. Yamaguchi et al., Phys. Rev. C **92**, 015805 (2015)
[9] K.P. Artemov, O.P. Belyanin, A.L. Vetoshkin, R. Wolskj, M.S. Golovkov, V.Z. Gol'dberg, M. Madeja, V.V. Pankratov, I.N. Serikov, V.A. Timofeev et al., Sov. J. Nucl. Phys **52**, 408 (1990)
[10] H. Yamaguchi, Y. Wakabayashi, S. Kubono, G. Amadio, H. Fujikawa, T. Teranishi, A. Saito, J. He, S. Nishimura, Y. Togano et al., Phys. Lett. B **672**, 230 (2009)
[11] J.J. He, L.Y. Zhang, A. Parikh, S.W. Xu, H. Yamaguchi, D. Kahl, S. Kubono, J. Hu, P. Ma, S.Z. Chen et al., Phys. Rev. C **88**, 012801 (2013)
[12] H. Yamaguchi, T. Hashimoto, S. Hayakawa, D.N. Binh, D. Kahl, S. Kubono, Y. Wakabayashi, T. Kawabata, T. Teranishi, Phys. Rev. C **83**, 034306 (2011)
[13] D. Kahl, H. Yamaguchi, S. Kubono, A.A. Chen, A. Parikh, D.N. Binh, J. Chen, S. Cherubini, N.N. Duy, T. Hashimoto et al., Phys. Rev. C **97**, 015802 (2018)
[14] S. Kubono, Y. Yanagisawa, T. Teranishi, S. Kato, T. Kishida, S. Michimasa, Y. Ohshiro, S. Shimoura, K. Ue, S. Watanabe et al., Eur. Phys. J. **A13**, 217 (2002)
[15] Y. Yanagisawa, S. Kubono, T. Teranishi, K. Ue, S. Michimasa, M. Notani, J.J. He, Y. Ohshiro, S. Shimoura, S. Watanabe et al., Nucl. Instrum. Meth. Phys. Res., Sect. A **539**, 74 (2005)
[16] A. Kim, N.H. Lee, M.H. Han, J.S. Yoo, K.I. Hahn, H. Yamaguchi, D.N. Binh, T. Hashimoto, S. Hayakawa, D. Kahl et al., Phys. Rev. C **92**, 035801 (2015)
[17] S. Hayakawa, S. Kubono, D. Kahl, H. Yamaguchi, D.N. Binh, T. Hashimoto, Y. Wakabayashi, J.J. He, N. Iwasa, S. Kato et al., Phys. Rev. C **93**, 065802 (2016)
[18] M. Wiescher, J. Görres, S. Graff, L. Buchmann, F.K. Thielemann, Astrophys. J. **343**, 352 (1989)
[19] C. Fröhlich, P. Hauser, M. Liebendörfer, G. Martínez-Pinedo, F.K. Thielemann, E. Bravo, N. Zinner, W. Hix, K. Langanke, A. Mezzacappa et al., Astrophys. J. **637**, 415 (2006)
[20] S. Wanajo, H.T. Janka, S. Kubono, Astrophys. J. **729**, 46 (2011)
[21] M. Freer, N.L. Achouri, C. Angulo, N.I. Ashwood, D.W. Bardayan, S. Brown, W.N. Catford, K.A. Chipps, N. Curtis, P. Demaret et al., Phys. Rev. C **85**, 014304 (2012)
[22] C. Angulo *et al.*, Nucl. Phys. A **656**, 3 (1999)
[23] Y. Xu, K. Takahashi, S. Goriely, M. Arnould, M. Ohta, H. Utsunomiya, Nucl. Phys. A **918**, 61 (2013)
[24] R.K. Wallace, S.E. Woosley, Astrophys. J. Suppl. Ser. **45**, 389 (1981)
[25] P.D. Ingalls, J.S. Schweitzer, B.D. Anderson, M. Rios, Phys. Rev. C **13**, 524 (1976)
[26] T. Hashimoto, H. Ishiyama, T. Ishikawa, T. Kawamura, K. Nakai, Y. Watanabe, H. Miyatake, M. Tanaka, Y. Fuchi, N. Yoshikawa et al., Nuclear Instruments and Methods in Physics Research Section A **556**, 339 (2006)
[27] R. Cyburt, A. Amthor, A. Heger, E. Johnson, L. Keek, Z. Meisel, H. Schatz, K. Smith, The Astrophysical Journal **830**, 55 (2016)

Proton-Induced Reactions of Astrophysical Interest

Thomas Chillery[1,⋆] *for the LUNA Collaboration*[2]

[1] *SUPA, School of Physics and Astronomy, University of Edinburgh, Edinburgh, UK*
[2] *http://luna.lngs.infn.it*

Abstract. A discrepancy exists between the ^{6}Li abundances predicted from big bang nucleosynthesis models and those measured in pre-main sequence stars. To further constrain the predicted abundances of ^{6}Li in these stars, high accuracy measurements are required of reactions destroying ^{6}Li. Namely ^{6}Li(p,γ)^{7}Be and ^{6}Li(p,α)^{3}He. These have recently been studied at the Laboratory for Underground Nuclear Astrophysics (LUNA) to measure their low energy cross sections. I present both the campaign's experimental setup and current status of the data analysis.

1 Introduction

Following the recent sucess of the Wilkinson Microwave Anisotropy Probe (WMAP) experiment in measuring the footprint of the cosmic microwave background (CMB) [1], cosmologists are closer than ever to precisely model and understand big bang nucleosynthesis (BBN) [2] in the early universe. Based on the current Standard Model, BBN aims to predict the abundances of light elements between one second and three minutes after the big bang. Of high interest to nuclear astrophysicists, the parameters obtained from BBN+WMAP data have been used to predict the present abundances of low mass isotopes found in stars. Whilst deuterium and ^{4}He abundances show excellent agreement between those predicted by BBN+WMAP and those measured in galaxies [2], large discrepancies are found for ^{6}Li and ^{7}Li between BBN+WMAP predictions and measurements of low-metallicity, or pre-main sequence (PMS), stars. A recent review [3] suggests the predicted ^{7}Li abundances are 2-4 times higher than those measured, and inversely the predicted ^{6}Li abundances are 5000 times lower than measured. Whilst this large discrepancy is currently attributed to ^{6}Li production via proton induced spallation and fusion reactions, the environments encountered in PMS stars are expected to destroy more ^{6}Li than ^{7}Li. To further constrain the predicted abundances of ^{6}Li in PMS stars, further high accuracy measurements are required of the reactions burning ^{6}Li. Understanding the destructive processes of ^{6}Li to a higher accuracy will also enable theorists to better model the inner profile of PMS stars. This profile can provide an indication of how the convective currents vary beneath the stellar surface [4].

Two specific ^{6}Li destruction reactions, ^{6}Li(p,γ)^{7}Be and ^{6}Li(p,α)^{3}He, have been studied at the underground 400 kV LUNA accelerator in Gran Sasso laboratory, Italy. The underground conditions lead to low background fluxes, thus experiments at energies of astrophysical interest are less challenging than similar studies performed on Earth's surface. Data collection was recently completed whilst

⋆e-mail: Thomas.Chillery@ed.ac.uk

analysis is currently ongoing. Recent work by [5] shows a very distinctive drop in the ^{6}Li(p,γ)^{7}Be S-factor for E_{cm} approaching the Gamow energy ($\simeq$ 15 keV). An upwards trend at decreasing E_{cm} would have been expected as suggested by earlier papers, see [6–9], and therefore [5] proposes a resonance exists with properties $E_R \simeq 195$ keV, $\Gamma_p \simeq 50$ keV to account for the drop. Whilst [5] does not quote whether their reported S-factor includes the electron screening effect or not (bare), the maximum possible enhancement may be calculated from [10]. At the lowest energy measured by [5], E_{cm} = 46.6 keV, this enhancement factor between bare and screened cross section is 3.1 %. This enhancement factor is smaller than the error in S-factor of 12.7 % reported by [5], suggesting the electron screening effect is negligible in the current literature. Determining the ^{6}Li(p,γ)^{7}Be S-factor at E_{cm} = 64 - 338 keV, thus confirming or denying the proposed resonance, is the final objective of this ongoing experimental campaign at LUNA.

2 Experimental Setup

A schematic diagram of the LUNA 400 kV accelerator solid target chamber used for the ^{6}Li runs is shown in figure 1. The accelerator was operated at energies E_{lab} = 75 - 394 keV (E_{cm} = 64 - 338 keV). The protons were focused on target using two circular apertures (diameters of 6 mm and 3 mm respectively) reducing the beamspot to a size comparable to the target (diameter 23 mm). The protons are then sent through a copper tube, which was cooled using LN_2 for carbon suppression and biased to -300 V to suppress secondary electrons. The proton beam strikes the solid target mounted at 55° to the beam axis. Alpha particles emitted during the experiment were detected using a silicon detector, mounted on an actuator for free movement inside the target chamber and facing the front of the target. The Si detector was collimated (collimator diameter 1 mm) and had a mylar foil (thickness 5 μm) in front of it to attenuate the backscattered proton beam. Gamma rays emitted were detected by the HPGe, placed on top of an aluminium slide and facing perpendicular to the rear of the target holder. The signals aquired from both detectors were shaped with standard NIM electronics and aquired using MAESTRO [11].

The silicon detector was calibrated using both a fixed ^{241}Am source and the well-known 151keV resonance in the ^{18}O(p,α)^{15}N reaction [12]. The HPGe detector was calibrated using fixed sources (^{137}Cs, ^{60}Co, and ^{88}Y) and the well-known 259keV resonance in the ^{14}N(p,γ)^{15}O reaction [13]. Various ^{6}Li-enriched targets were used during the two months of experiment; $^6Li_2O$ with nominal thicknesses 20-40 μg/cm^2, $^6LiWO_4$ with nominal thicknesses 100-130 μg/cm^2, and an "infinitely thick" (>450 μg/cm^2) LiCl target. The oxide and tungstate targets were evaporated on tantalum backings, while the LiCl has a Cu backing. The targets were water cooled during beam bombardment.

3 Preliminary Results

To calibrate the setup we measured the fixed sources, ^{14}N(p,γ)^{15}O, and ^{18}O(p,α)^{15}N reactions mentioned in section 2. The HPGe efficiency curves for a range of distances, d=4-254 mm, is presented in figure 2, where the distance is measured between the front cap of the Germanium and the back side of the target holder. The efficiency errors plotted are statistical, of order 1.5 %. Systematic uncertainties arising from correct positioning of the HPGe detector are estimated to be no greater than 5 %. The efficiency curves do not include true summing corrections. A sample spectrum obtained from protons of E_p = 230 keV incident on a 40 μg/cm^2 $^6Li_2O_4$ target is shown in figure 2. The direct capture peaks for the decay of the compound nucleus to the ground state, γ_1, and to the 429 keV excited state, γ_2, are labelled. The first and second escape peaks are also observed. A beam induced reaction caused by contaminants in the target chamber, ^{19}F(p, $\alpha\gamma$)^{16}O, was also monitored online and is shown in the same figure.

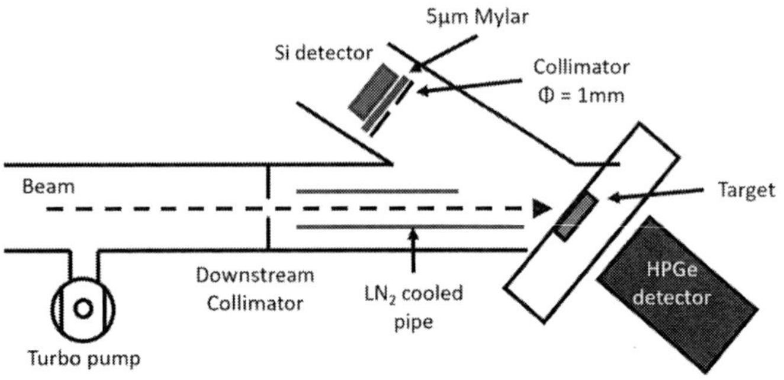

Figure 1: A schematic diagram of the target chamber, with Silicon and Germanium detector positions shown.

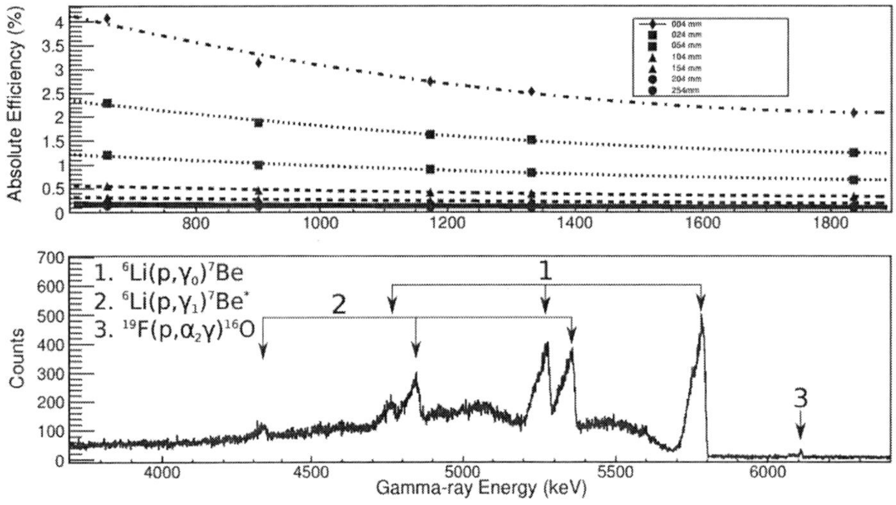

Figure 2: Top: Preliminary HPGe Efficiency Curves. Errors plotted are statistical (1.5 %) but smaller than data point sizes. Bottom: Gamma spectrum for protons of E_p = 230 keV on $^6Li_2O_4$ target.

4 Summary

During 2017 an experimental campaign was recently completed studying two destructive reactions; ^{6}Li(p,γ)^{7}Be and ^{6}Li(p,α)^{3}He, in the LUNA laboratory. Multiple ^{6}Li-enriched solid targets were irradiated using a proton beam at E_{cm} = 64 - 338 keV. Efficiency curves for both the HPGe and Si detectors have been obtained using both fixed sources and proton beam-induced reactions. Gamma detection efficiency errors are 1.5 % statistical, 5 % systematic. True summing corrections for the HPGe efficiency, including the use of Geant4 simulations, are currently ongoing. Efforts to extract the S-factor from the beam studies for both destructive reactions is also ongoing.

References

[1] C. Bennett et al., The Astrophysical Journal Supplement Series **208**, 20 (2013)
[2] B. Fields, P. Molaro, S. Sarkar, Chinese Physics C pp. 1–15 (2014)
[3] B. Fields, Annual Review of Nuclear and Particle Science **61**, 47 (2011)
[4] F. Thévenin et al., Astronomy & Astrophysics **598**, A64 (2017)
[5] J. He et al., Physics Letters B **725**, 287 (2013)
[6] Z. Switkowski et al., Nuclear Physics A **331**, 50 (1979)
[7] F. Barker, Australian Journal of Physics **33**, 159 (1980)
[8] K. Arai, D. Baye, P. Descouvemont, Nuclear Physics A **699**, 963 (2002)
[9] J. Huang, C. Bertulani, V. Guimarães, Atomic Data and Nuclear Data Tables **96**, 824 (2010)
[10] H. Assenbaum, K. Langanke, C. Rolfs, Z. Phys. A. Atomic Nuclei **327**, 461 (1987)
[11] *Maestro multichannel analyser emulation*, http://www.ortec-online.com/products/application-software/maestro-mca
[12] C. Bruno et al., Eur. Phys. J. A **51**, 94 (2015)
[13] G. Imbriani et al., Eur. Phys. J. A **25**, 455 (2005)

Improved information on astrophysical *S*-factor for the ^{10}B(p,α_0)^{7}Be reaction using the Trojan Horse method

A. Cvetinović[1,2], *C.* Spitaleri[1,2], *R.* Spartá[1,2], *G.G.* Rapisarda[1,2], *S.M.R.* Puglia[1,2], *M.* La Cognata[1], *S.* Cherubini[1,2], *G.L.* Guardo[1], *M.* Gulino[1,3], *L.* Lamia[1,2], *R.G.* Pizzone[1], *S.* Romano[1,2], *M.L.* Sergi[1], and *A.* Tumino[1,3]

[1] *INFN-Laboratori Nazionali del Sud, Catania, Italy*
[2] *Dipartimento di Fisica e Astronomia, University of Catania, Catania, Italy*
[3] *Facoltà di Ingegneria e Architettura, Università degli Studi di Enna "Kore", Enna, Italy*

Abstract. The ^{10}B(p,α_0)^{7}Be reaction has been studied by applying the Trojan Horse method to the ^{2}H(^{10}B,α_0 ^{7}Be)n reaction. The bare-nucleus astrophysical $S(E)$-factor in absolute units was extracted in a wide energy range, from 2.2 MeV to 3 keV and normalized to the direct experimental data, thus allowing determination of the electron screening potential for which a value of U_e=391±74 eV was obtained.

1 Introduction

For a proper understanding of the stellar structure and evolution, the precise determination of the abundances of light elements such as Li, Be and B is crucial. Since the depletion of these elements occurs at different depths in stars, their surface abundances are strongly influenced by the burning process as well as by the extension of the convective layer. Therefore, these elements can be used as a probe for the internal stellar structure [1–3]. The ^{10}B burning process mostly proceeds via the (p,α) reaction. The cross section for this process is dominated by a strong resonance at 10 keV ($J^\pi=\frac{5}{2}^+$, 8.699 MeV ^{11}C level), laying exactly at the energy corresponding to the Gamow peak (E_G) [4]. Since the presence of low-energy resonances can introduce significant uncertainties in the extrapolation procedure, the proper evaluation of the ^{10}B(p,α)^{7}Be reaction rate can be done by applying the Trojan Horse Method (THM) [5–13] to the ^{2}H(^{10}B,α ^{7}Be)n three-body reaction, escaping in this way all the difficulties related to the Coulomb barrier suppression or electron screening effects. The experiment presented in this paper is a further study of the ^{10}B(p,α_0)^{7}Be reaction, performed in order to complete the previously started experimental program [14, 15].

2 Experimental set-up

In order to study the ^{10}B(p,α_0)^{7}Be reaction, the experiment has been performed at the *INFN - Laboratori Nazionali del Sud* in Catania. A 56 μg/cm^2 self-supported deuterated polyethylene (CD_2) target was bombarded with the 28 MeV ^{10}B ion beam, intensity of around 2 enA, provided by the SMP Tandem Van de Graaff accelerator. Using a collimator, the beam spot on the target was reduced to 1 mm in diameter. The detection system (sketched in Fig. 1) has been consisted of four 500 μm thick Position

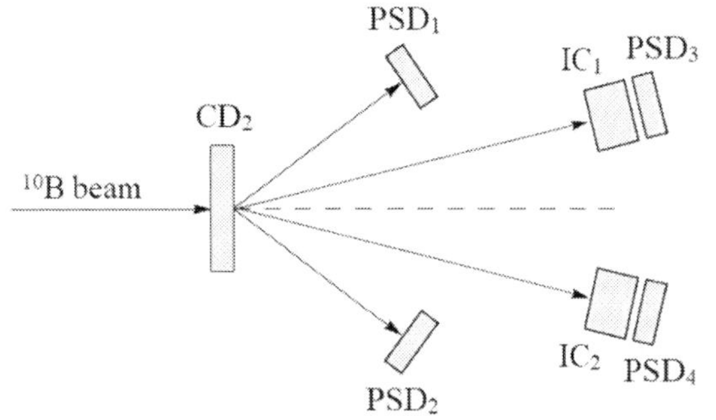

Figure 1. A schematic view of the experimental set-up.

Sensitive Silicon Detectors (PSD) two of which were centred at opposite sides of the beam axis at the laboratory angle of 14.5° covering an angular range of ±3.5°. Other two PSD's were fixed at the laboratory angle of -28°/+28° and covered an angular range of ±7°. Particle identification was carried out using the standard ΔE-E technique. Namely, two Ionization Chambers (IC) filled with isobutane gas (C_4H_{10}) at a pressure of 100 mbar with 1.5 μm thick Mylar entrance and exit windows, have been placed in front of an inner pair of PSD's (PSD_3 and PSD_4 in Fig. 1). A symmetric configuration of the experimental apparatus has been chosen in order to double the statistic. The deterioration of the target was continuously controlled by monitoring the ratio between the measured ^{7}Be yield and the charge collected in the Faraday cup at the end of the beam line.

3 Data analysis and results

After the position-energy calibration and selection of the events corresponding to the investigated three-body reaction were done, following the standard THM procedure, the selected experimental data were compared with the Monte Carlo simulation, confirming the correct identification of the ^{2}H(^{10}B,α_0 ^{7}Be)n reaction (a detailed description of the procedure is given in [14, 15]). Further, with the aim to deduce the two-body cross section by applying the THM formalism, a restriction for neutron momentum of -40 MeV/c $\leq p_s \leq$ 40 MeV/c, deduced by studying the shape of the spectator (neutron) momentum distribution, was set in order to select the Quasifree (QF) contribution and to minimize the other reaction mechanisms that can possibly produce the same particles in the final state (^{7}Be, α and n), such as Sequential Decay (SD) or Direct Breakup (DBU).

Since the Plane-Wave Impulse Approximation (PWIA) was used for the calculation of the two-body cross section $\sigma(E)$, this quantity was obtained in arbitrary units [13] and a normalization of TH data to the ones obtained from direct measurements was necessary. However, due to different direct data sets providing cross section functions that are not consistent between each other [14–24], to obtain the bare-nucleus TH S-factor in absolute units, the THM data have been normalised to the R-matrix calculation of [15], leading to a more accurate S_b(10 keV)=2950±291 MeV b value comparing to S_b(10 keV)=2942±395 MeV b reported in [15]. Fig. 2 shows a result obtained for the astrophysical S-factor as a function of center-of-mass energy in comparison with data from [15] and data from [14] renormalised as suggested in [15]. As it can be seen, all three sets of data are in good agreement within their error bars.

Due to an optimization of the experimental set-up we improved energy resolution by 80% in the present experiment compared to the previous THM study [15], thus ensuring more accurate determination of the ^{10}B(p,α_0)^{7}Be bare-nucleus astrophysical S-factor and consequently, of the electron

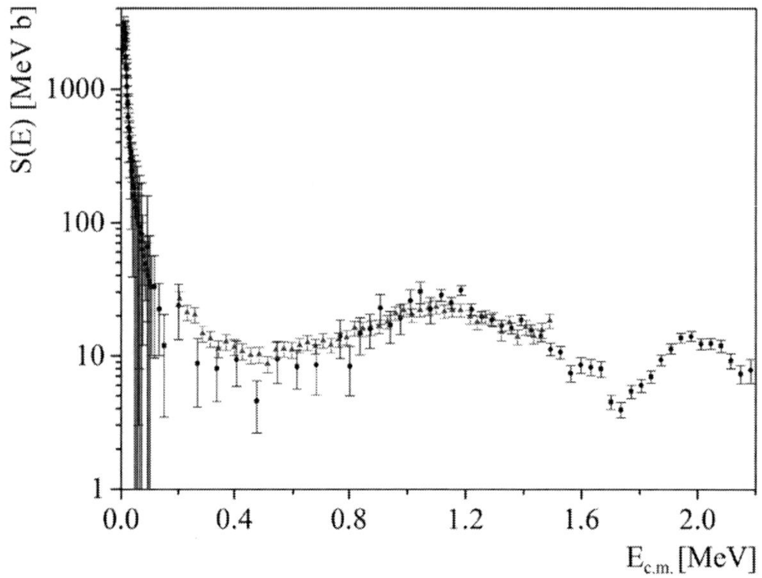

Figure 2. TH astrophysical *S(E)*-factor (black points) compared with the data from [15] (red triangles) and data from [14] (blue squares) renormalised as suggested in [15].

screening potential (U_e). This was also possible due to the wide energy range available for the normalization procedure. A value for the electron screening potential of U_e=391±74 eV has been determined and found to be in a good agreement (within uncertainties) with the theoretical value calculated for the adiabatic limit U_e^{ad}=340 eV.

References

[1] A. M. Boesgaard et al., Astrophys. J. **621**, 991 (2005)
[2] L. Lamia et al., Astrophys. J. **768**, 65 (2013)
[3] L. Lamia et al., Astrophys. J. **811**, 99 (2015)
[4] C. Rolfs and W. S. Rodney, *Cauldrons in the Cosmos* (The University of Chicago Press, Chicago, 1988)
[5] G. Baur, Phys. Lett. B **178**, 135, (1986)
[6] C. Spitaleri, *in Problems of Fundamental Modern Physics, II*, (World Scientific, Singapore, 1990), p. 21.
[7] S. Typel and H. H. Wolter, Few-Body Syst. **29**, 75 (2000)
[8] R. E. Tribble et al., Rep. Progr. Phys. **77**, 106901 (2014)
[9] S. Cherubini et al., Phys. Rev. C **92**, 015805 (2015)
[10] R. G. Pizzone et al., Eur. Phys. J. A **52**, 24 (2016)
[11] R. G. Pizzone et al., Astrophys. J. **836**, 57 (2017)
[12] M. Gulino et al., Phys. Rev. C **87**, 012801(R) (2013)
[13] C. Spitaleri et al., Phys. At. Nucl. **74**, 1725 (2011)
[14] C. Spitaleri et al., Phys. Rev. C **90**, 035801 (2014)
[15] C. Spitaleri et al., Phys. Rev. C **95**, 035801 (2017)

[16] J. Szabó et al., Nucl. Phys. A **195**, 527 (1972)
[17] N. A. Roughton et al., At. Data Nucl. Data Tables **23**, 177 (1979)
[18] M. Wiescher et al., Phys. Rev. C **28**, 1431 (1983)
[19] M. Youn, Nucl. Phys. A **533**, 321 (1991)
[20] C. Angulo et al., Z. Phys. A **345**, 231 (1993)
[21] T. Rauscher and G. Raimann, Phys. Rev. C **53**, 2496 (1996)
[22] I. Lombardo et al., J. Phys. G: Nucl. Part. Phys. **43**, 045109 (2016)
[23] A. Caciolli, Eur. Phys. J. A **52**, 136 (2016)
[24] M. Wiescher et al., Phys. Rev. C **95**, 044617 (2017)

The ^{19}F(α, p)^{22}Ne and ^{23}Na(p,α)^{20}Ne reaction in AGB nucleosynthesis via THM

G. D'Agata[1,2], *R. G.* Pizzone[1], *M.* La Cognata[1], *I.* Indelicato[1], *C.* Spitaleri[1,2], *V.* Burjan[3], *S.* Cherubini[1,2], *A.* Di Pietro[1], *G. L.* Guardo[4], *M.* Gulino[1,5], *M.* La Commara[6,7], *L.* Lamia[1], *M.* Lattuada[1,2], *M.* Mazzocco[8,9], *J.* Mrazek[3], *M.* Milin[10], *S.* Palmerini[11,12], *C.* Parascandolo[6], *D.* Pierroutsakou[6], *G.G.* Rapisarda[1], *S.* Romano[1,2], *M.L.* Sergi[1], *N.* Soič[10], *R.* Spartá[1], *O.* Trippella[11,12], and *A.* Tumino[1,5]

[1] *INFN - Laboratori Nazionali del Sud, Via Santa Sofia 62, 95123, Catania, Italy*
[2] *Dipartimento di Fisica e Astronomia, Università degli Studi di Catania, Via S. Sofia, 64, 95123, Catania, Italy*
[3] *Institute for Nuclear Physics, Prague-Rez, Czech Republic*
[4] *ELI-NP, Str. Reactorului no.30, P.O.BOX MG-6, Bucharest - Magurele, Romania*
[5] *Facoltà di Ingegneria ed Architettura, Kore University, Viale delle Olimpiadi, 1, I-94100 Enna, Italy*
[6] *INFN - Sezione di Napoli, via Cintia, 80125 Napoli, Italy*
[7] *Dipartimento di Scienze Fisiche, Università "Federico II", via Cintia, 80125 Napoli, Italy*
[8] *Dipartimento di Fisica e Astronomia, Università di Padova, Via Marzolo 8, I-35131 Padova, Italy*
[9] *INFN - Padova, Via Marzolo 8, I-35131 Padova, Italy*
[10] *Department of Physics, University of Zagreb, Bijenicka 32, Zagreb, Croatia*
[11] *Dipartimento di Fisica e Geologia, Università degli Studi di Perugia, via A. Pascoli s/n, 06123, Perugia, Italy*
[12] *INFN-sezione di Perugia, via A. Pascoli s/n, 06123, Perugia, Italy*

Abstract. In AGB environment, fluorine and sodium abundances are still matter of debate. About ^{19}F (only stable isotope of fluorine), its abundance in the universe is strictly related to standard and extra-mixing processes taking place inside AGB-stars, that are considered to be the most important sites for its production. Nevertheless the way in which it is destroyed is far from being well understood. On the other hand, ^{23}Na presence in Globular Clusters, along with is well-known anticorrelation with oxygen has made clear that this element must be produced in previous generations stars, and intermediate-mass AGB stars are one of the possible candidates for its production. For this reason we studied the ^{19}F(α,p)^{22}Ne and ^{23}Na(p,α)^{20}Ne reactions in the energy range of relevance for astrophysics via the Trojan Horse Method (THM), using the three-body reactions ^{6}Li(^{19}F,p^{22}Ne)d and ^{23}Na(d,pn)^{20}Ne.

1 Introduction

The only Fluorine stable isotope - ^{19}F - has been clearly observed in AGB stars [1], where it can be produced in the He-intershell region through the chain of reactions ^{18}O(p,α)^{15}Na(α,γ)^{19}F. Due to the high abundance of alpha particles in the He-intershell, the ^{19}F(α,p)^{22}Ne reaction is expected to dominate over the competitive reaction channels in this region. Fluorine can in fact be destroyed by

the ^{19}F(n,γ)^{20}F reaction, triggered by the neutrons produced in ^{13}C(α,n)^{16}O and ^{22}Ne(α,n)^{25}Mg reactions, and via the ^{19}F(p,α)^{16}O (already studied using THM by [2] and [3]) and ^{19}F(α,p)^{22}Ne. It is now clear how Fluorine abundance is really sensitive to the physical condition of the stars, and can be used as a probe to clarify if stellar interior nucleosynthesis is well understood or not [4, 5]: in this case Fluorine abundance can not be reproduced by the up-to-date models. A possible reason of this fact is the large uncertaities at Helium burning temperatures ($0.2 \leq T_9 \leq 0.8$), due to the lack of experimental data about the cross-section in the energy region of astrophysical interest. In particular, before this measurement (results partially published in [6, 7]), there were no experimental data below 0.660 MeV in the center-of-mass reference frame, being the Gamow window between 0.200 and 1.200 MeV. This is caused by the presence of the Coulomb barrier. In cases like this one, the use indirect methods such as the Trojan Horse Method (THM) [8] can be rather useful to overcome the technical difficulties related to the height of the Coulomb barrier.
Another reaction of great interest in AGB nucleosynthesis is the ^{23}Na(p, α)^{20}Ne, that is cosidered to have great importance in intermediate-mass AGB stars ($M = 4 \div 8\ M_\odot$), and could be strongly related to the wide known Na/O anticorrelation in globular clusters [10]. This reaction is also of great importance because it represents the turning point between the NeNa and MgAl cycles. Both have the result to fuse hydrogen into helium, and for a mass number $20 \leq A \leq 40$, both (p, α) and (p, γ) channels are open, so those kind of reaction will compete. One of the two cycles can be active if the reaction rate branching ratio $B_{p\alpha/p\gamma} = N_A\langle\sigma v\rangle_{p\alpha}/N_A\langle\sigma v\rangle_{p\gamma}$ is large enough. H−burning in the mass range $A \geq 20$ is important to understand Ne, Na, Mg and Al abundances observed in stars: the relative isotopic abundance depends on the temperature and density conditions inside the H−burning region of a certain star. About NeNa−cycle, at temperature $T \sim 6 \cdot 10^6$ K, ^{22}Ne is entirely transformed in ^{23}Na. An extra production of this element is predicted at temperatures higher than $35 \cdot 10^6$ K, reaching 60% at $T \sim 60 \cdot 10^6$ K. This extra production is provided by ^{20}Ne reaction. In the end ^{23}Na starts burning at $T \geq 60 \cdot 10^6$ K [11].
^{23}Na(p, α)^{20}Ne has not been studied at astrophysical energies with direct methods in the energy range of astrophysical interest. Here the Gamow window lies bwetween 50 keV and 200 keV, while the Coulomb barrier is at 2.57 MeV. Several states of ^{24}Mg were however studied [12], via the ^{23}Na(^{3}He, d)^{24}Mg transfer reaction at 20 MeV. Two resonant states at 37 keV and 138 keV were found: the former had a too low cross section to be studied (but uncertainties were reduced by a factor of 515), and the latter is still the bigger source of uncertainties (circa a factor of 12) in the temperature region near $T \sim 70 \cdot 10^6$ K.

2 The Experiment

To study the ^{19}F(α,p)^{22}Ne reaction, the THM was applied by using a ^{6}Li beam (6 MeV energy, 5 enA intensity) impinging on a ^{7}LiF target (150 μm thick). The ^{6}Li particle can be considered as a cluster $\alpha \oplus d$. The aim of this experiment was to induce the ^{6}Li(^{19}F,p^{22}Ne)d three-body reaction to study the ^{19}F(α,p)^{22}Ne two-body one: here the α particle is considered as the participant to the two-body one, while the deuteron continues its course undisturbed, and is therefore considered the spectator for the process of interest. Following the THM prescriptions, the beam energy was chosen to measure the ^{19}F(α,p)^{22}Ne cross-section in the energy region of interest for astrophysics.
The experiment was performed at Ruđer Bošković Institute (Zagreb, Croatia) and the set-up (Fig. 1) was composed by two ΔE-E telescopes (placed at 12.3°±7° and 32.3°±7°) made up using thin silicon detectors (82 mm^2 active surface, 15μm and 9μm thickness respectively) as ΔE stage and thick (500μm each) Position Sensitive Detectors (PSDs) as E stage, both meant for deuteron particles detection. On the opposite side of the beam, another 3 PSDs with the same specifics are placed at 37.7°±12°, 81°±9° and 119.9°±11° and are devoted to proton detection. The necessity to have detectors sensible to the

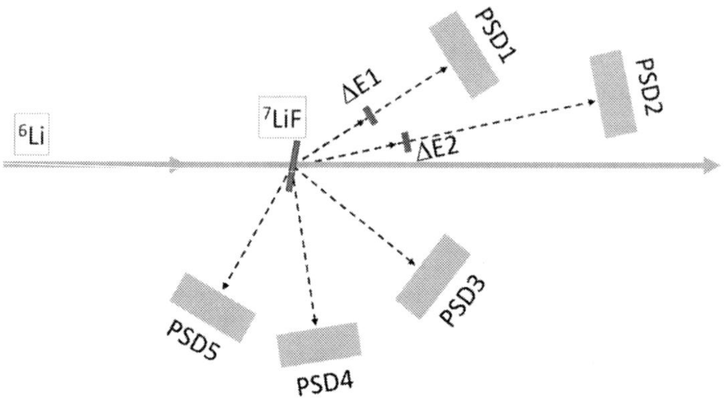

Figure 1. Sketch of the adopted experimental set-up. The two ΔE-E telescopes are situated on one side of the beam axis. Three PSDs were placed on the other side with respect to the beam direction.

position is given by the fact that one of the three particles in the exit channel of the $^{6}Li(^{19}F,p^{22}Ne)d$, ^{22}Ne, is not experimentally detected, but its characteristics are reconstructed form the energies and positions of the two measured ones.
About the THM application in this case the Trojan Horse (TH) nucleus is in the beam and we chose to reveal the spectator (deuterons) along with the light outgoin particle (protons). This choice was made considering the characteristics of the ^{22}Ne particles coming from the $^{19}F(\alpha,p)^{22}Ne$ reaction, that are emitted at very forward angles ($\vartheta_{Ne} \leq 6°$) - where the elastic scattering $^{6}Li + ^{19}F$ (circa 4kHz) would eventually break down the detectors - and at energies (3.5 MeV) that are not high enough to emerge from the target.
In order to maximize the statistics on PSD1 and PSD2, it was also decided to tilt the target at 10° with respect to the beam direction in order to maximize deuteron detection. This is crucial because of the low reaction rate for the reaction and the low energy ($E_d \leq$ 5MeV, determined by means of a proper Monte Carlo simulation) of the deuteron particles.
About the $^{23}Na(p,\alpha)^{20}Ne$ reaction, the THM was applied using a brand new ^{23}Na beam (58 MeV energy, 0.8 enA intensity) delivered at Laboratori Nazionali del Sud. This beam impinged on a CD_2 target, inducing the $^{23}Na(d,pn)^{20}Ne$ three-body reaction: here the Trojan Horse nucleus is in the terget. Deuteron in fact can be thought as a cluster composed by a proton and a neutron: the first particle participates to the reaction, inducing the $^{23}Na(p,\alpha)^{20}Ne$, while the second acts as a spectator. The experimental set-up consisted in two symmetrical ΔE-E telescopes - composed by a ionization chamber filled with isobuthane (50 mb internal pressure) and two PSDs (500 μm thick) - centered at 6° and two thick PSDs (1000 μm) at higher angles (centered at 25 °). The telescopes were used to identify and measure ^{22}Ne energy and angle of emission, while the thick PSDs were meant for alpha particles.

3 Results and Conclusions

3.1 The $^{19}F(\alpha,p)^{22}Ne$ reaction

Once deuterons are identified by means of the ΔE-E technique, the Q-value for the three-body process can be isolated and the quasi-free contribution can be separated from the sequential decay (procedure explained in [7]). The half-of-energy-shell binary cross-section of interest can than be written, in quasi-free conditions and in the simplest approach, as a function of the measured three-body differential cross-section as follows:

$$\left(\frac{d\sigma}{d\Omega}\right)^{HOES} \propto \left(KF\,|\Phi(p_s)|^2\right)^{-1} \cdot \frac{d^3\sigma}{dE_{CM}d\Omega_d\Omega_p} \tag{1}$$

Here KF is a kinematic factor, and $|\Phi(p_s)|^2$ is tied to the relative motion of the TH nucleus: p_s in fact represents the momentum of the spectator particle (deuteron) inside the TH nucleus (^{6}Li). Dividing the measured three-body triple differential cross-section for KF and $|\Phi(p_s)|^2$ it is therefore possible to obtain the two-body one of interest in arbitrary units. Those data were then fitted by means of the *Modified R-Matrix* procedure [2], and a first evaluation of the widths of the involved resonances has been performed [6, 7]. The measured cross-section was then normalized to direct data [9] in the overlap region (0.6÷0.9 MeV in the center-of-mass reference frame), thus obtaining the absolute units cross-section for the ^{19}F(α,p)^{22}Ne reaction. A comparison between the existing reaction rate [9] and the new one was also attempted, and it resulted in an enhancement - in the temperature window of astrophysical interest - up to a factor of four for the destruction rate of ^{19}F via the (α,p) reaction (Tab. 1).

Temperature [10^9K]	Rate $\left[\frac{cm^3}{mol \times sec}\right]$	$\frac{R_{THM}}{R}$
0.10	2.59×10^{-22}	1.08
0.20	6.00×10^{-14}	1.81
0.30	4.67×10^{-10}	3.71
0.40	1.11×10^{-7}	3.29
0.50	5.09×10^{-6}	1.66
0.60	1.14×10^{-4}	1.12

Table 1. Reaction rate and ratio between experimental THM results (R_{THM}) and the one parametrized from the results of [9] (R), $\frac{R_{THM}}{R}$, for several temperatures in units of 10^9 K

In conclusion, for the first time the ^{19}F(α,p)^{22}Ne cross-section inside the energy range of astrophysical relevance was measured, and its impact on the reaction rate has been evaluated. An investigation on the effect of this new rate on fluorine nucleosynthesis is still in progress [7].

3.2 The ^{23}Na(p,α)^{20}Ne reaction

After the selection of the heavy fragment (^{20}Ne), again the Q-value for the three-body process can be isolated and the quasi-free contribution can be separated from the sequential decay for both coincidences between a ΔE-E telescopes and the 1000 μm thick PSD placed on the opposite side of the beam (Fig.2). Once the channel is selected, an evaluation of the momentum distribution of the participant particle inside the Trojan Horse nucleus has been made. Its experimental behaviour follows a Hultén function (FWHM≃52 MeV/c), in agreement with literature [13]. The procedure reported in section 3.1 can therefore be used. In the future, the half-of-energy-shell binary cross-section of interest will be extracted and the impact of this new measurement on the reaction rate will be evaluated.

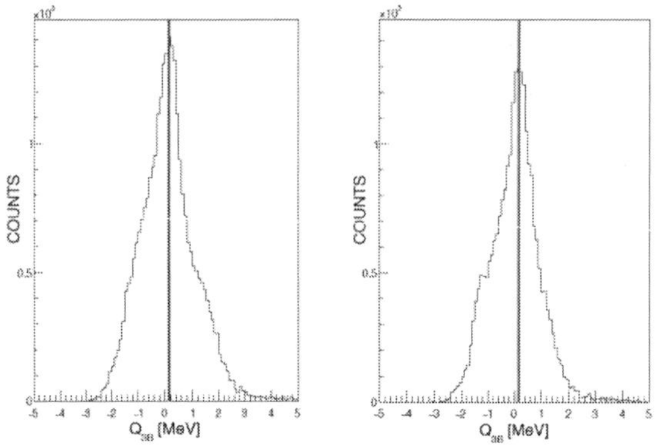

Figure 2. Q_{value} of the three body reaction for both coincidences: the red line indicates the theoretic value for the three body reaction (Q_{value}=0.152 MeV)

This publication was supported by OP RDE, MEYS Czech Republic under the project EF16_013/0001679

References

[1] Lucatello, S., Masseron, T., Johnson, J. et al. 2011, ApJ, **729**, 40
[2] La Cognata M., Mukhamezhanov A., Spitaleri C., et al., 2011, ApJ L, **739**, L 54
[3] Indelicato, I., La Cognata, M., Spitaleri, C., et al., 2017, ApJ, **845**, 1
[4] Lugaro, M., Ugalde, C., Karakas, A. I., et al. 2004, ApJ, **615**, 934
[5] Cristallo, S., Di Leva, A., Imbriani, G. et al. 2014, A&A, **570**, 46
[6] Pizzone, R. G., D'Agata, G., La Cognata M., Indelicato I., et al. 2017, ApJ, **836**, 57
[7] D'Agata, G., Pizzone, R. G., La Cognata M., Indelicato I., et al., ApJ, **submitted**
[8] Spitaleri, C., La Cognata, M., Lamia, L., et al. 2016, Eur. Phys. J. A, **52**, 77
[9] Ugalde, C., Azuma, R. E., Coutre, A., et al. 2008, Phys. Rev. C, **77**, 035801
[10] Ventura, P. and D'Antona, F., 2011, MNRAS, **410**, 2760-2766
[11] Mowlavi, N., 1999, A&A, **344**, 617-631
[12] Hale, S.E., Champagne, A.E., Iliadis, C., et al 2004, Phys.Rev.C, **70** , 4
[13] Pizzone, R. G., Spitaleri, C., Mukhamedzhanov, A. M., et al., 2009, Phys. Rev. C, **80**, 025807

^{7}Be(n,p) cross section measurement for the Cosmological Lithium Problem at the n_TOF facility at CERN

Lucia Anna Damone[1,*], *N.* Colonna[1], *M.* Barbagallo[1,2], *M.* Mastromarco[1], *J.* Andrzejewski[3], *P.* Finocchiaro[4], *L.* Cosentino[4], and ***The n_TOF Collaboration***

[1] *INFN, Bari*
[2] *CERN*
[3] *University of Lodz, Poland*
[4] *INFN Laboratori Nazionali del Sud, Catania, Italy*

Abstract. One of the most puzzling problems in Nuclear Astrophysics is the "Cosmological Lithium Problem", i.e the discrepancy between the primordial abundance of ^{7}Li observed in metal poor halo stars [1], and the one predicted by Big Bang Nucleosynthesis (BBN). One of the reactions that could have an impact on the problem is ^{7}Be(n,p)^{7}Li. Despite of the importance of this reaction in BBN, the cross-section has never been directly measured at the energies of interest for BBN. Taking advantage of the innovative features of the second experimental area at the n_TOF facility at CERN, an accurate measurement of ^{7}Be(n,p) cross section has been recently performed at n_TOF, with a pure ^{7}Be target produced by implantation of a ^{7}Be beam at ISOLDE. The experimental procedure, the setup used in the measurement and the results obtained so far will be here presented.

1 Introduction

Theoretical models of the Big Bang Nucleosynthesis (BBN), i.e. the production of light elements in the first few minutes of the life of the Universe, correctly predict the abundance of all primordial elements except for ^{7}Li that is overestimated by a factor of 3-5. This significant discrepancy between observation and predictions is known as the cosmological lithium problem (CLiP) [2]. Since 95% of the primordial ^{7}Li is produced by the electron capture decay of ^{7}Be, a higher destruction rate of ^{7}Be can solve or at least partially explain the CLiP. In this scenario, reactions induced by neutrons on ^{7}Be, in particular the ^{7}Be(n,α)^{4}He and the ^{7}Be(n,p)^{7}Li reactions, could play an important role in explaining this discrepancy. However, data on these reactions have been so far scarce or completely missing. The recent construction of a second experimental area (EAR2) at n_TOF (Neutron Time of Flight) characterized by an extremely high instantaneous neutron flux (10^{8} n/cm^{2}/pulse), a good energy resolution and a low repetition rate, offered the unique opportunity to perform time-of-flight measurements of ^{7}Be(n,p)^{7}Li and ^{7}Be(n,α)^{4}He cross sections over a wide energy range (from thermal up to 1 GeV), covering the one of interest for the Big Bang Nucleosynthesis. Results on the latter reaction have already been published [3]. Preliminary results on the (n,p) reaction are here reported.

2 Experimental Setup

At n_TOF the neutron beam is produced by spallation of a pulsed proton beam from the CERN Proton Synchrotron accelerator (PS), with momentum of 20 GeV/c, impinging on a cylindrical lead target surrounded by water for cooling and neutron moderation purposes. For the first 13 years of operation, only one experimental area was available (EAR1) [4]. In 2014, a new experimental hall located on the vertical direction at 20 m distance from the spallation target, the so-called n_TOF-EAR2 [5][6] was completed and became operational. The main advantage of this new measuring station with respect

*e-mail: lucia.damone@ba.infn.it

to the existing one consists in a flux up to 40 times higher. Combined with the shorter time-of-flight at a given energy, 10 times lower than EAR1 due to the shorter flight-path, the higher flux results in more than two orders of magnitude higher signal-to-background ratio for radioactive samples. These features of the EAR2 neutron beam open the way to measurements of neutron-induced reactions on very thin samples, reactions of low cross sections or radioactive isotopes with short half-life such as the ^{7}Be (53 days). The measurement of the ^{7}Be(n,p)^{7}Li reaction was performed using a telescope made of two Silicon strip detectors placed off the neutron beam track (see Figure 1) [7]. The system was calibrated by means of the well known ^{6}Li(n,α)^{3}H reaction, whose cross-section is standard from thermal energy up to 1 MeV.

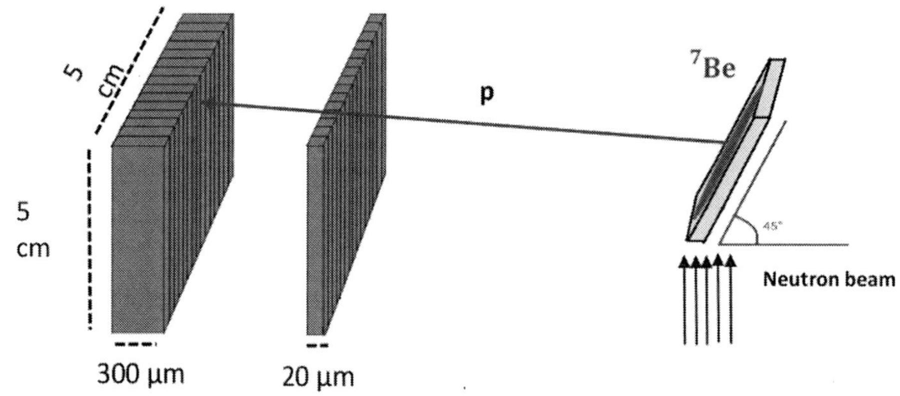

Figure 1: Schematic view of the of the experimental setup. In yellow the support for ^{7}Be, 1.5x1.5 cm^{2}; in red the ^{7}Be deposit of 2.5 mm radius. The Silicon strip detectors are in blue. The ΔE is 20 μm thick, while E 300 μm.

Two different ^{7}Be samples of 20 MBq and 1.1 GBq respectively were prepared starting from 200 GBq ^{7}Be solution collected at Paul Scherrer Institute (PSI). The samples were produced at ISOLDE by implantation of a beam of ^{7}Be at 35 keV on an Aluminum backing.

3 Data analysis

Once the samples were prepared, they were placed in the neutron beam line at EAR2 inside the vacuum chamber hosting a silicon telescope. Protons emitted in the ^{7}Be(n,p)^{7}Li reaction were detected and identified, as shown in the E-ΔE matrix of the events detected in the telescope (Figure 2). The red circle indicates the region of protons from the ^{7}Be(n,p)^{7}Li reaction.

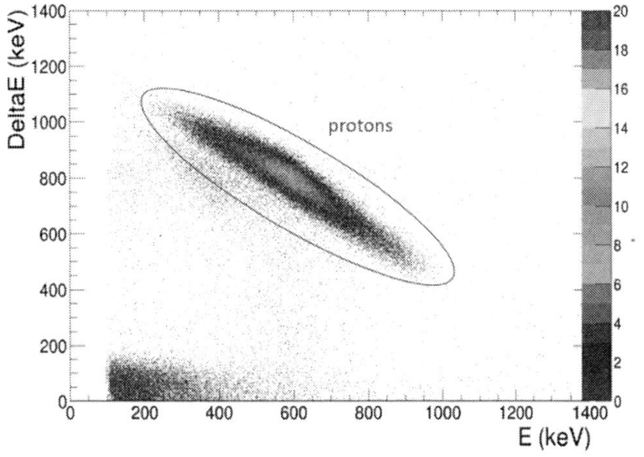

Figure 2: Coincidence of protons between the strips of the two array of Silicon detectors.

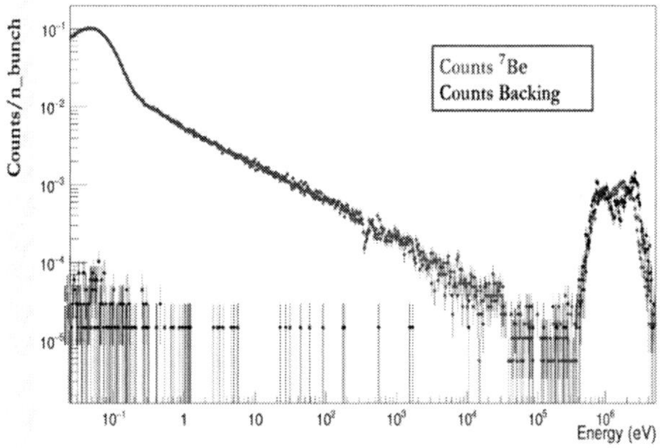

Figure 3: Comparison between the counts collected with the ^{7}Be sample, in red, and the dummy aluminum target, in black.

We did also a study of the background, obtained by comparing the counts collected with the ^{7}Be sample and the ones from the backing. The signal/background ratio is good up to a few hundred keV, while above this energy the background related to the ^{14}N(n,p) reaction dominates (Figure 3). In order to calibrate the detector and make a reference measurement, prior to the ^{7}Be(n,p)^{7}Li measurement, the ^{6}Li(n,α)^{3}H reaction, whose cross section is standard, was measured and the result compared with the evaluation from ENDF/B-VII.1 [7], as shown in Figure 4. A good agreement, within 5%, from thermal neutron energy up to 1 MeV has been observed.

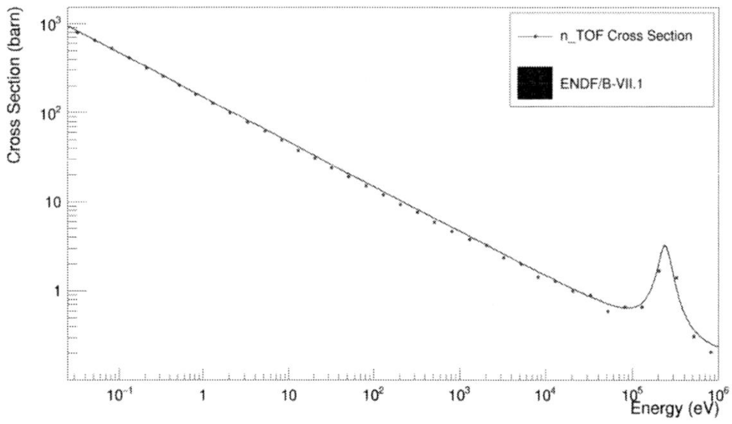

Figure 4: Comparison between our measured ^{6}Li(n,α)^{3}H cross section and the evaluation from ENDF/B-VII.1 data library.

4 Conclusions

The ^{7}Be(n,p)^{7}Li cross section measurement has been performed for the first time covering the energy range of interest for the Cosmological Lithium Problem (20-200 keV) in the second experimental area (EAR2) at n_TOF, using 1.1 GBq and 20 MBq pure ^{7}Be samples implanted at the General Low Mass (GLM) beam line of the ISOLDE facility, starting from a 200 GBq ^{7}Be solution collected at PSI. The ^{7}Be(n,p)^{7}Li cross-section has been measured with the two different samples and the results are in a very good agreement. Analysis is in progress to extract the absolute value and energy dependence of this important reaction, and determine its influence on the CLiP problem.

References

[1] Martin Asplund et al. *LITHIUM ISOTOPIC ABUNDANCES IN METAL-POOR HALO STARS.* The Astrophysical Journal, 644:229–259, 2006 June 10.

[2] Brian D. Fields *The Primordial Lithium Problem.* Departments of Astronomy and of Physics, University of Illinois, Urbana, Illinois 61801;

[3] M. Barbagallo et al., *The ^{7}Be(n,α)^{4}He reaction and the Cosmological Lithium Problem: measurement of the cross section in a wide energy range at n_TOF (CERN).* Phys. Rev. Lett. 117, 152701 (2016)

[4] C. Guerrero et al., *Performance of the neutron time-of-flight facility n_TOF at CERN,* Eur.Phys.J.A **49**, 27 (2013)

[5] M. Sabatè-Gilarte et al., *High-accuracy determination of the neutron flux in the new experimental area n_TOF-EAR2 at CERN* Eur. Phys. J. A (2017), **53**: 210

[6] C.Weiss et al., *The new vertical neutron beam line at CERN n_TOF facility design and outlook on the performance,* NIM A (2015) **799**: 90.

[7] M. Barbagallo et al., *Experimental setup and procedure for the measurement of the ^{7}Be(n,p)^{7}Li reaction at n_TOF.* Nucl. Instr. Meth. A (2018), **887**: 27-33.

[8] M.B. Chadwick, M. Herman, P. Obložinský, M.E. Dunn, Y. Danon, A.C. Kahler, D.L. Smith, B. Pritychenko, G. Arbanas, R. Arcilla, R. Brewer, D.A. Brown, R. Capote, A.D. Carlson, Y.S.

Cho, H. Derrien, K. Guber, G.M. Hale, S. Hoblit, S. Holloway, T.D. Johnson, T. Kawano, B.C. Kiedrowski, H. Kim, S. Kunieda, N.M. Larson, L. Leal, J.P. Lestone, R.C. Little, E.A. McCutchan, R.E. MacFarlane, M. MacInnes, C.M. Mattoon, R.D. McKnight, S.F. Mughabghab, G.P.A. Nobre, G. Palmiotti, A. Palumbo, M.T. Pigni, V.G. Pronyaev, R.O. Sayer, A.A. Sonzogni, N.C. Summers, P. Talou, I.J. Thompson, A. Trkov, R.L. Vogt, S.C. van der Marck, A. Wallner, M.C. White, D. Wiarda, P.G. Young. "ENDF/B-VII.1 Nuclear Data for Science and Technology: Cross Sections, Covariances, Fission Product Yields and Decay Data", *Nuclear Data Sheets*, **1**12, pp. 2887 - 2996, 2011.

First Measurement of $^{72}\mathrm{Ge}(n,\gamma)$ at n_TOF

M. Dietz[1,★], C. Lederer-Woods[1], O. Aberle[2], J. Andrzejewski[3], L. Audouin[4], M. Bacak[5,2,6], J. Balibrea[7], M. Barbagallo[8], F. Bečvář[9], E. Berthoumieux[6], J. Billowes[10], D. Bosnar[11], A. Brown[12], M. Caamaño[13], F. Calviño[14], M. Calviani[2], D. Cano-Ott[7], R. Cardella[2], A. Casanovas[14], F. Cerutti[2], Y. H. Chen[4], E. Chiaveri[2,10,15], N. Colonna[8], G. Cortés[14], M. A. Cortés-Giraldo[15], L. Cosentino[16], L. A. Damone[8,17], M. Diakaki[6], C. Domingo-Pardo[18], R. Dressler[19], E. Dupont[6], I. Durán[13], B. Fernández-Domínguez[13], A. Ferrari[2], P. Ferreira[20], P. Finocchiaro[16], V. Furman[21], K. Göbel[22], A. R. García[7], R. Garg[1], A. Gawlik[3], S. Gilardoni[2], T. Glodariu[23], I. F. Gonçalves[20], E. González-Romero[7], E. Griesmayer[5], C. Guerrero[15], F. Gunsing[6,2], H. Harada[24], S. Heinitz[19], J. Heyse[25], D. G. Jenkins[12], E. Jericha[5], F. Käppeler[26], Y. Kadi[2], D. Kahl[1], A. Kalamara[27], P. Kavrigin[5], A. Kimura[24], N. Kivel[19], M. Kokkoris[27], M. Krtička[9], D. Kurtulgil[22], E. Leal-Cidoncha[13], H. Leeb[5], J. Lerendegui-Marco[15], S. Lo Meo[28,29], S. J. Lonsdale[1], D. Macina[2], J. Marganiec[3,30], T. Martínez[7], A. Masi[2], C. Massimi[29,31], P. Mastinu[32], M. Mastromarco[8], E. A. Maugeri[19], A. Mazzone[8,33], E. Mendoza[7], A. Mengoni[28], P. M. Milazzo[34], F. Mingrone[2], A. Musumarra[16,35], A. Negret[23], R. Nolte[30], A. Oprea[23], N. Patronis[36], A. Pavlik[37], J. Perkowski[3], I. Porras[38], J. Praena[38], J. M. Quesada[15], D. Radeck[30], T. Rauscher[39,40], R. Reifarth[22], C. Rubbia[2], J. A. Ryan[10], M. Sabaté-Gilarte[2,15], A. Saxena[41], P. Schillebeeckx[25], D. Schumann[19], P. Sedyshev[21], A. G. Smith[10], N. V. Sosnin[10], A. Stamatopoulos[27], G. Tagliente[8], J. L. Tain[18], A. Tarifeño-Saldivia[14], L. Tassan-Got[4], S. Valenta[9], G. Vannini[29,31], V. Variale[8], P. Vaz[20], A. Ventura[29], V. Vlachoudis[2], R. Vlastou[27], A. Wallner[42], S. Warren[10], C. Weiss[5], P. J. Woods[1], T. Wright[10], P. Žugec[11,2], and the n_TOF Collaboration

[1]*School of Physics and Astronomy, University of Edinburgh, United Kingdom*
[2]*European Organization for Nuclear Research (CERN), Switzerland*
[3]*University of Lodz, Poland*
[4]*Institut de Physique Nucléaire, CNRS-IN2P3, Univ. Paris-Sud, Université Paris-Saclay, F-91406 Orsay Cedex, France*
[5]*Technische Universität Wien, Austria*
[6]*CEA Irfu, Université Paris-Saclay, F-91191 Gif-sur-Yvette, France*
[7]*Centro de Investigaciones Energéticas Medioambientales y Tecnológicas (CIEMAT), Spain*
[8]*Istituto Nazionale di Fisica Nucleare, Sezione di Bari, Italy*
[9]*Charles University, Prague, Czech Republic*
[10]*University of Manchester, United Kingdom*
[11]*Department of Physics, Faculty of Science, University of Zagreb, Zagreb, Croatia*
[12]*University of York, United Kingdom*
[13]*University of Santiago de Compostela, Spain*
[14]*Universitat Politècnica de Catalunya, Spain*
[15]*Universidad de Sevilla, Spain*
[16]*INFN Laboratori Nazionali del Sud, Catania, Italy*

★e-mail: mirco.dietz@ed.ac.uk

[17] *Dipartimento di Fisica, Università degli Studi di Bari, Italy*
[18] *Instituto de Física Corpuscular, Universidad de Valencia, Spain*
[19] *Paul Scherrer Institut (PSI), Villingen, Switzerland*
[20] *Instituto Superior Técnico, Lisbon, Portugal*
[21] *Joint Institute for Nuclear Research (JINR), Dubna, Russia*
[22] *Goethe University Frankfurt, Germany*
[23] *Horia Hulubei National Institute of Physics and Nuclear Engineering, Romania*
[24] *Japan Atomic Energy Agency (JAEA), Tokai-mura, Japan*
[25] *European Commission, Joint Research Centre, Geel, Retieseweg 111, B-2440 Geel, Belgium*
[26] *Karlsruhe Institute of Technology, Campus North, IKP, 76021 Karlsruhe, Germany*
[27] *National Technical University of Athens, Greece*
[28] *Agenzia nazionale per le nuove tecnologie (ENEA), Bologna, Italy*
[29] *Istituto Nazionale di Fisica Nucleare, Sezione di Bologna, Italy*
[30] *Physikalisch-Technische Bundesanstalt (PTB), Bundesallee 100, 38116 Braunschweig, Germany*
[31] *Dipartimento di Fisica e Astronomia, Università di Bologna, Italy*
[32] *Istituto Nazionale di Fisica Nucleare, Sezione di Legnaro, Italy*
[33] *Consiglio Nazionale delle Ricerche, Bari, Italy*
[34] *Istituto Nazionale di Fisica Nucleare, Sezione di Trieste, Italy*
[35] *Dipartimento di Fisica e Astronomia, Università di Catania, Italy*
[36] *University of Ioannina, Greece*
[37] *University of Vienna, Faculty of Physics, Vienna, Austria*
[38] *University of Granada, Spain*
[39] *Department of Physics, University of Basel, Switzerland*
[40] *Centre for Astrophysics Research, University of Hertfordshire, United Kingdom*
[41] *Bhabha Atomic Research Centre (BARC), India*
[42] *Australian National University, Canberra, Australia*

Abstract. The slow neutron capture process (s-process) is responsible for producing about half of the elemental abundances heavier than iron in the universe. Neutron capture cross sections on stable isotopes are a key nuclear physics input for s-process studies. The ^{72}Ge(n,γ) cross section has an important influence on production of isotopes between Ge and Zr during s-process in massive stars and therefore experimental data are urgently required. ^{72}Ge(n,γ) was measured at the neutron time-of-flight facility n_TOF (CERN) for the first time at stellar energies. The measurement was performed using an enriched ^{72}GeO$_2$ sample at a flight path of 185 m with a set of liquid scintillation detectors (C_6D_6). The motivation, experiment and current status of the data analysis are reported.

1 Introduction

The main ideas of stellar nucleosynthesis were proposed around 60 years ago by Cameron [1] and Burbidge, Burbidge, Fowler and Hoyle [2], who introduced two neutron-driven processes for the formation of heavy elements between Fe and Bi. Half of these abundances are formed in the slow neutron capture process (**s-process**), which happens at low neutron density n_{n} of up to $10^{11}\ \frac{\mathrm{n}}{\mathrm{cm}^3}$, in which case faster radioactive β-decays guide the reaction path along the valley of stability. Neutrons are provided mainly by two source reactions: ^{13}C$(\alpha,n)^{16}$O and ^{22}Ne$(\alpha,n)^{25}$Mg. With understanding the abundance pattern, different components of the s-process were introduced.

The **main** component contributes to all abundances from Fe to Bi, but mainly isotopes between Zr and Bi. It occurs in low - and intermediate mass TP-AGB (thermally pulsing asymptotic giant branch) stars at temperatures of $(0.1-0.3)\cdot 10^9$ K. A sufficient neutron exposure establishes a reaction flow

equilibrium, which means that the product of cross section times abundance is constant (details can be found in Ref. [3]).
The **weak** component of the s-process, which dominantly contributes to abundances of isotopes up to $A = 90$ takes place in massive stars ($M > 8M_{\odot}$). The s-process occurs during convective helium-core burning around $T = 0.3 \cdot 10^9\,$K and in subsequent convective carbon-shell burning around $T = 1.0 \cdot 10^9\,$K.
Neutrons are thermalised in these hot stellar environments and follow a Maxwell-Boltzmann velocity distribution. The effective stellar neutron capture cross section for the s-process is therefore a Maxwellian Average Cross Section (MACS), at the corresponding neutron energies of $kT \approx 8, 25$ and 90 keV.

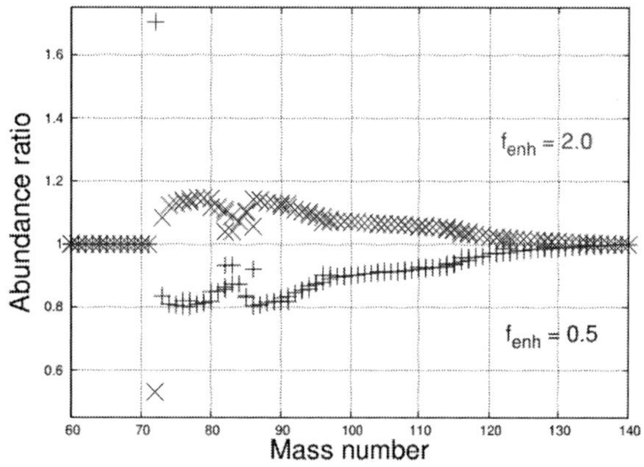

Figure 1:
NETZ-calculation [4] shows that only the change of one single MACS of ^{72}Ge by an enhancement factor $f_{\rm enh}$ of 2 or $\frac{1}{2}$ produces a change up to 50 to 70% for the isotope itself, but as well up to 20% changes for heavier isotopes up to $A = 125$.

In the weak s-process neutron exposures are not high enough to establish a reaction flow equilibrium. Therefore, an indivual MACS has influence not only on the abundance of its isotope, but also on abundances of isotopes following the reaction chain as well [5]. This propagation effect is shown for the case of ^{72}Ge in Fig. 1. Changes in the abundances up to ± 20% for heavier isotopes up to Te ($A = 125$) are observed for enhancement of the ^{72}Ge(n, γ)-MACS by factors of 2 and $\frac{1}{2}$, respectively. For ^{72}Ge(n, γ) only one measurement above thermal neutron energies exists [6], covering energies up to a few keV, while no experimental data [7] are available for the astrophysically important higher keV region. Therefore, experimental data are required for neutron energies up to about 200 keV to determine MACSs for the entire energy range of interest.

2 Measurement and Data Analysis

The measurement of the ^{72}Ge(n, γ) cross section [8] was performed at the neutron time-of-flight facility n_TOF, located at CERN. Details of n_TOF are described here [9]. At n_TOF, neutrons over a large energy range (25 meV to several GeV) are produced by spallation reactions of a highly energetic (20 GeV/c), pulsed proton beam (from the CERN Proton Synchrotron) impinging on a massive Pb target, yielding in an instantaneous neutron intensity of $\sim 2 \times 10^{15}$ neutrons per pulse.
The capture measurement was performed using a 96.59 %-enriched ^{72}GeO$_2$ sample at a distance of $L \approx 185$ m from the spallation target at Experimental Area 1 (EAR-1). The prompt γ rays, following a neutron capture event, were detected by a set of four liquid scintillation C_6D_6 detectors, which are optimized to have an extremely low sensitivity to neutrons scattered from the sample [10, 11]. A photograph of the setup is shown in Fig. 2a. For each capture event, the neutron energy is determined

by measuring the time-of-flight t_n, i.e. the time difference between production of the neutron and detection of the capture events, using Eq. 1:

$$E_n = m_n c^2 (\frac{1}{\sqrt{1-\beta^2}} - 1) \quad \text{with } \beta = \frac{L}{t_n c} \tag{1}$$

The counting spectrum can be transformed to the neutron capture yield using

$$\mathbf{Y} = f_N \frac{(\mathbf{C} - \mathbf{B})}{\Phi \cdot \varepsilon}, \tag{2}$$

where **C** is the number of counts measured with the $^{72}GeO_2$ sample and **B** is the background, which is measured with no sample in the beam. The detection efficiency ε depends on the de-excitation path of the compound nucleus. A detection efficiency independent of the latter can be achieved by applying the Pulse Height Weighting Technique [12], which requires detailed Monte-Carlo simulations of the detector response functions, taking into account the experimental geometry. The neutron flux spectrum Φ is measured with different detector arrays, e.g. a set of silicone detectors measuring alphas and tritons after neutron capture on a thin ^{6}Li-target. Finally a normalisation f_N to the 4.9 eV resonance in ^{197}Au(n,γ), which is saturated in the capture yield [13], is applied to determine the yield **Y**.
In the following, a preliminary counting spectrum is plotted for ^{72}Ge(n,γ) in Fig. 2b, showing a number of neutron resonances in the stellar energy range. In further analysis, resonances will be analysed from the neutron capture yield and used to determine the MACS.

3 Summary

The ^{72}Ge(n,γ) was measured for the first time at stellar energies at n_TOF (EAR-1). This cross section is of importance to determine abundances produced in the s-process in massive stars.

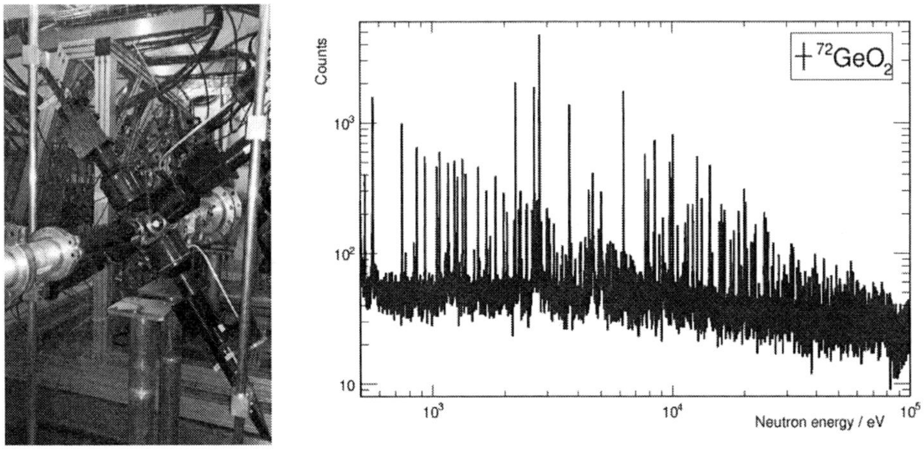

(a) (b)

Figure 2: (a) Neutron capture setup at EAR-1 shows four C_6D_6 detectors with the $^{72}GeO_2$ sample in the middle. The neutron beam is coming from the left. (b) Preliminary unweighted counting spectrum of neutron capture on $^{72}GeO_2$ is plotted in the neutron energy region of interest.

The experiment was performed using an enriched $^{72}GeO_2$ sample and a set of liquid scintillation detectors (C_6D_6) to detect prompt γ rays emitted after neutron capture. The analysis procedure has been shortly introduced and preliminary counting spectrum was shown. The data are still under analysis and first experimental results at stellar energies can be expected soon.

Acknowledgement

This work was supported by the Adolf Messer Foundation, by the Austrian Science Fund FWF (J 3503), the Science and Technology Facilities Council UK (ST/M006085/1), and the European Reseach Council ERC-2015-STG Nr. 677497.

References

[1] A. G. W. Cameron, Pub. Astron. Soc. Pacific **69**, 201-222 (1957)
[2] M. Burbidge et al., Rev. Mod. Phys. **29**, 547-650 (1957)
[3] F. Käppeler et al., Rev. Mod. Phys. **83**, 157-193 (2011)
[4] M. Weigand et al., Phys. Rev. C **92**, 045810 (2015)
[5] M. Pignatari et al., The Astroph. J. **710**, 1557-1577 (2010)
[6] K. Maletski et al., At. Energ. USSR **24**, 173 (1968)
[7] I. Dillmann et al., AIP Conference Proceedings **819**, 123-127 (2006)
[8] C. Lederer et al., Proposal to ISOLDE and neutron time-of-flight committee CERN-INTC-2013-021 , INTC-P-381 (2013)
[9] C. Guerrero et al., Eur. Phys. J. A **49**, 27 (2013)
[10] P. F. Mastinu et al., CERN-n_TOF-PUB-2013-002 (2013)
[11] R. Plag et al., Nucl. Instr. Meth. Phys. Res. A **496**, 425-436 (2003)
[12] U. Abbondanno et al., Nucl. Instr. Meth. Phys. Res. A **521**, 454-467 (2004)
[13] R. L. Macklin et al., Nucl. Instr. Meth. **164**, 213 (1979)

Development of the ELISSA array: prototype testing at Laboratori Nazionali del Sud

G. L. Guardo[1,2,⋆], *A.* Anzalone[2], *D.* Balabanski[1], *S.* Chesnevskaya[1], *W.* Crucillá[2], *D.* Filipescu[1], *M.* Gulino[2,3], *M.* La Cognata[2], *D.* Lattuada[1], *C.* Matei[1], *R. G.* Pizzone[2], *G. G.* Rapisarda[2], *S.* Romano[2,4], *C.* Spitaleri[2,4], *A.* Taffara[2], *A.* Tumino[2,3], and *Y.* Xu[2]

[1] *ELI-NP/ IFIN-HH, Magurele, Romania*
[2] *INFN, Laboratori Nazionali del Sud, Catania, Italy*
[3] *Facoltá di Ingegneria e Architettura, Universitá Kore, Enna, Italy*
[4] *Dipartimento di Fisica e Astronomia, Universitá degli Studi, Catania, Italy*

Abstract. The Extreme Light Infrastructure-Nuclear Physics (ELI-NP) facility, under construction in Magurele near Bucharest in Romania, will provide high-intensity and high-resolution gamma ray beams that can be used to address hotly debated problems in nuclear astrophysics, such as the accurate measurements of the cross sections of the ^{24}Mg(γ,α)^{20}Ne reaction
For this purpose, a silicon strip detector array (named ELISSA) will be realized in a common effort by ELI-NP and Laboratori Nazionali del Sud (INFN-LNS), in order to measure excitation functions and angular distributions over a wide energy and angular range. A prototype of ELISSA was built and tested at INFN-LNS in Catania (Italy) with the support of ELI-NP. In this occasion, we have carried out experiments with alpha sources and with a 11 MeV ^{7}Li beam that show up a very good energy resolution (better than 1%) and very good position resolution, of the order of 1 mm. Moreover, a threshold of 150 keV can be easily achieved with no cooling.

1 Introduction

The upcoming ELI-NP facility consists of two major components: the High Power Laser System and the Gamma Beam System (GBS). ELI-NP will allow either combined or stand-alone experiments using the high-power laser and the gamma beam [1]. The ELI-NP GBS will provide, for the first time, pencil size gamma beams in the range between 200 keV and 19.5 MeV with a bandwidth better than 0.5%, spectral density of about 10^4 photons/s/eV and linear polarization higher than 95% [2].
Such a facility will open new experimental perspectives for studies in the field of photonuclear physics. Moreover, thanks to these excellent features, the ELI-NP GBS will provides unique opportunities to respond to the need of Nuclear Astrophysics to perform accurate measurements of small cross sections (order of μb or even less) of nuclear reactions of the hydrogen and helium burning processes and hence of the astrophysical S-factors that are essential for stellar evolution modeling. As an example, a direct ^{24}Mg photodissociation measurement using gamma beams of energies 10-12 MeV will allow us to

⋆e-mail: luca.guardo@eli-np.ro

determine a much more accurate cross section to be used in nuclear reaction network calculations to improve the knowledge of the pre-supernova chemical composition.
Indeed, silicon burning sets the chemical composition of the star right before the core collapse and the subsequent supernova explosion, thus constituting a key process for the understanding of core-collapse supernovae [3]. In this framework, the ^{24}Mg(γ,α)^{20}Ne reaction governs the downward flow from ^{24}Mg to ^{4}He, thus determining the effective rate of ^{28}Si destruction, making its reaction rate critically important to stellar models [3]. At present, the ^{24}Mg(γ,α)^{20}Ne reaction rate has been calculated from the ^{20}Ne(α,γ)^{24}Mg rate. In the temperature range of interest, around $3.9 \cdot 10^9$ K, the ^{20}Ne(α,γ)^{24}Mg reaction rate may be subject to systematic errors of the order of a factor of 2, as can be seen from the different results reported in [4]. For those reasons, a new measurement using the gamma radiation beam with unique characteristics at ELI-NP may allow us to resolve conflicting data.

2 The ELISSA detector

Position sensitive silicon strip detectors represent a good solution for detecting charged particles as they provide high position resolution over a large solid angle with a relatively limited channel count. Moreover, a silicon array would make it possible to measure reactions on solid targets with the possibility to measure excitation functions and angular distributions over a wide range allowing for a better understating of reaction mechanism. Resistive strip silicon detector arrays have been successfully designed and commissioned for studies of nuclear astrophysics reactions, e g. ORRUBA (Oak Ridge Rutgers University Barrel Array) or ANASEN (Array for Nuclear Astrophysics Studies with Exotic Nuclei) [5, 6].
In the case of photonuclear reactions of astrophysical relevance, since photodissociations are induced at photon energies slightly larger than particle emission thresholds owing to the typical temperatures in stars, the emitted fragments have low energies, ranging from few hundreds keV to few MeV. Therefore, low-threshold detectors are necessary. Thus, the Extreme Light Infrastructure Silicon Strip Array (ELISSA) is under construction [7]. The performed Monte Carlo simulation using a code based on GEANT4 tracking libraries and the n-body event generator of ROOT, as described in [8], proved that the barrel configuration is particularly suited as it guarantees a very good resolution and granularity, ensuring also a compact detection system (useful as it would allow a simple integration with ancillary detectors, such as neutron arrays) and a limited number of electronics channel [9]. The final setup of the ELISSA array will consist of X3 silicon-strip detectors (manufactured by Micron Semiconductor Ltd.) arranged into a barrel configuration that could be made up of 3 rings of 12 position sensitive detectors, for a total angular coverage of 100° in the laboratory system. The angular coverage is extended by using end cap detectors such as the assembly of four QQQ3 segmented detectors by Micron Semiconductor [10].

3 Test on ELISSA prototype

To test the performance of the X3 detectors and make sure that their energy and angular resolutions are suitable for the realization of the ELISSA detector, a detector prototype has been constructed at INFN-LNS Catania (Italy) in collaboration with the ELI-NP group (Romania). The prototype is made up of a single X3 detector and a single QQQ3 detector, so it replicates in a smaller scale the whole detector, and their readout is performed by means of standard analog electronics. In this occasion, we have carried out experiments with alpha sources and with a 11 MeV ^{7}Li beam at the INFN-LNS tandem. To measure the detection threshold, we have performed a run using a standard 3-peak alpha source and a ^{241}Am source shielded by a 17 μm thick Al foil. This degrader shifted the energy peak

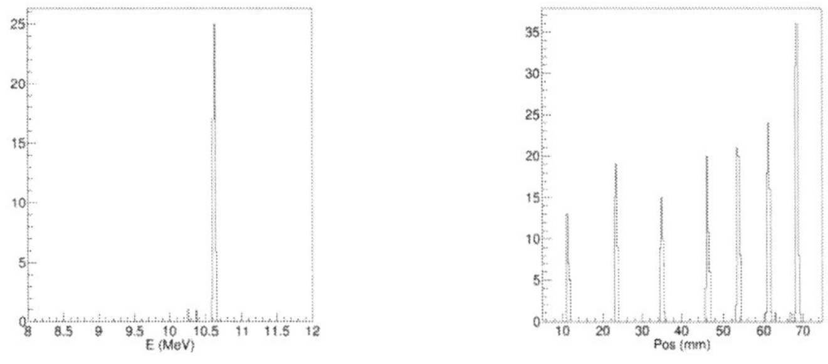

Figure 1. Left panel: Energy spectrum for a fixed position of ^{7}Li scattered off gold. Right panel: Position spectrum for the scattering induced by the ^{7}Li beam on Au target.

to 1 MeV and, due to energy straggling, the energy range spanned reached zero. In this way, we determined the energy threshold under standard conditions (no cooling) to be 150 keV. Energy and position resolutions were measured detecting scattering and reactions induced by the ^{7}Li beam on Au, C and CD_2 targets. In this way, we determined an energy resolution of $\leq 1\%$ for the case of scattering off gold (left panel of figure 1). At energies of about 11 MeV, the position resolution is better than 1 mm (right panel of figure 1) while at lower energies, below about 1 MeV, a position resolution of about 6 mm was found [11]. In this case, the worst position resolution is still good enough for the measurement of angular distribution since it correspond to an angular resolution of about 1.5°.
The good preliminary results of our tests allows us to say that the X3 detectors, as well the standard QQQ3 detectors for which the position resolution is fixed by the strip size and energy resolution is constant to a good approximation, are perfectly suited for nuclear astrophysics studies with ELISSA.

Acknowledgments

This work is supported in part by the Extreme Light Infrastructure Nuclear Physics (ELI-NP) Phase II, a project cofinanced by the Romanian Government and the European Union through the European Regional Development Fund - the Competitiveness Operational Programme (1/07.07.2016, COP, ID 1334).

References

[1] Zamfir N. V., EPJ Web of Conf. **66**, 11043 (2014)
[2] D. L. Balabanski et al., EPL **117**, 28001 (2017)
[3] C. Iliadis, *Nuclear Physics of Stars*, (Weinheim, Wiley-VCH Verlag, 2007)
[4] C. Angulo et al., Nucl. Phys. **A 656**, 3 (1999)
[5] S. D. Pain et al., Nucl. Instr. Meth. Phys. Res. **B 261**, 1122 (2007)
[6] M. Matos et al., PoS **NIC XI**, 226 (2010)
[7] M. La Cognata et al., JINST **12**, C03079, (2017)
[8] D. Lattuada et al. EPJ Web of Conf. **165**, 01034 (2017)
[9] O. Tesileanu et al., Rom. Rep. in Phys. **68**, S699 (2016)
[10] Micron Semiconductor, General cata logue, http://www.micronsemiconductor.co.uk/pdf/cat.pdf
[11] G. L. Guardo et al. EPJ Web of Conf. **165**, 01026 (2017)

Study of the contribution of the ^{7}Be(d, p) reaction to the ^{7}Li problem in the Big-Bang Nucleosynthesis

A. Inoue[1,*], *A.* Tamii[1], *K.* Abe[2], *S.* Adachi[1], *N.* Aoi[1], *M.* Asai[3], *M.* Fukuda[4], *G.* Gey[1], *T.* Hashimoto[5], *E.* Ideguchi[1], *J.* Isaak[1], *N.* Kobayashi[1], *Y.* Maeda[6], *H.* Makii[3], *K.* Matsuta[4], *M.* Mihara[4], *M.* Miura[1], *T.* Shima[1], *H* Shimizu[2], *R.* Tang[1], *T. Dinh* Trong[1], *H.* Yamaguchi[2], and *L.* Yang[2]

[1] *Research Center for Nuclear Physics, Osaka University*
[2] *Center for Nuclear Study, University of Tokyo*
[3] *Japan Atomic Energy Agency*
[4] *Department of Physics, Osaka University*
[5] *Institute for Basic Science*
[6] *Department of Applied Physics, University of Miyazaki*

Abstract. Our research goal is to measure the ^{7}Be(d, p) reaction to shed light on the ^{7}Li problem in the Big-Bang Nucleosynthesis. We are developing an unstable ^{7}Be target for a high-resolution measurement of the ^{7}Be(d, p)^{8}Be reaction. We plan to compare two methods to produce the ^{7}Be target: (1) *Activation method*, and (2) *Implantation method*. We performed an activation method experiment at the Van de Graaff at Osaka University, and obtained the cross-section data. A second experiment to obtain more accurate data will take place at the Tandem Electrostatic Accelerator, Kobe University. We have also made a ^{7}Be target with implantation method at CRIB, Center for Nuclear Study, University of Tokyo. An experiment to measure the (d, p) reaction with the implanted target is scheduled for 2018 at Japan Atomic Energy Agency, tandem facility.

1 Introduction

The overestimation of primordial ^{7}Li abundance in the standard Big-Bang Nucleosynthesis (BBN) model is one of the known and unresolved problem. A recent theoretical BBN model predicts a primordial ^{7}Li abundance that is about 3 times larger than the recent precise observation [**?**]. The difference is quite large while the abundance of the other light nuclei are reproduced well. This is one of the biggest problems in the BBN models and it illustrates the incomplete knowledge of the processes of the primordial formation of our universe.

Light nuclei were produced up to ^{7}Be by nuclear reactions in several hundred seconds following the Big Bang (Figure **??**). ^{7}Li nuclei were predominantly produced by the β decay of ^{7}Be in the standard BBN model. The β decay half life of ^{7}Be, 53.22 days = 5 million seconds, is much longer

*e-mail: azusa@rcnp.osaka-u.ac.jp

than the timescale of the production of light nuclei after the Big Bang. Thus, one possible scenario to solve the ^{7}Li problem is that ^{7}Be was destructed in the timescale of the nuclear reactions.

There are several possibilities to destroy ^{7}Be, for example the ^{7}Be(d, p)^{8}Be, the ^{7}Be(n, α) or ^{7}Be(n, p) reactions are considered [**?**]. We focus on the ^{7}Be(d, p)^{8}Be reaction since the contribution from ^{7}Be(d, p)^{8}Be is suggested to be larger than ^{7}Be(n, α)^{4}He [**? ?**].

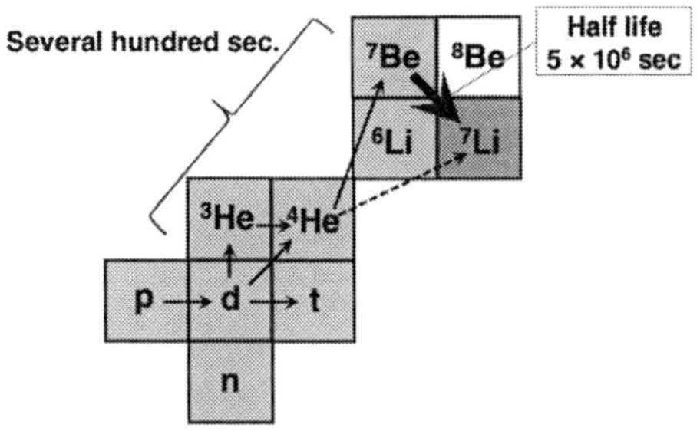

Figure 1. Light nuclei were produced in several hundred seconds following the Big Bang. ^{7}Li was produced by the β decay of ^{7}Be. The β decay half life of ^{7}Be is 5 million seconds, which is much longer than the timescale of the nuclear reactions.

2 Experimental Method and development

The goal of the experiment is to measure the cross section of the ^{7}Be(d, p)^{8}Be reaction in the BBN energy region of 100 - 400 keV. We plan to measure the ^{7}Be(d, p)^{8}Be reaction in direct kinematics with a ^{7}Be target since the available data are insufficient in the accuracy or in the energy range [**? ?**]. We are developing an unstable ^{7}Be target for a high-resolution measurement of the ^{7}Be(d, p)^{8}Be reaction. The development of the unstable target is a big technical challenge. We apply two different methods to make the ^{7}Be target, the *Activation method*, and the *Implantation method*.

2.1 Activation method

The activation method is a simple but effective way to make the ^{7}Be target. The first experiment was performed at the Van de Graaff, Osaka University, in December 2015. We used a natural Li target as a host target to produce the ^{7}Be target. A proton beam at 2.6 MeV was injected to activate ^{7}Li in the natural Li target to ^{7}Be. We immediately changed the beam to deuteron after the target production and measured proton from the (d, p) reaction by three layer Si detectors. Cu absorbers were placed in front of Si detectors to prevent low charged particles from damaging the detectors. The incident deuteron energy was 2 MeV. The analysis was done by the "thick target method". This method can be applied when the target is thick and the energy loss of incident beam is obvious in the target. The reacted incident deuteron energy was obtained by the detected proton energy. The number of the ^{7}Be target was calculated depending of the thickness of the Li host target. We obtained the cross section using the activation target. Figure **??** shows the measured cross sections of the ^{7}Be(d, p) reaction as a function of on the incident deuteron energy.

This cross section at 0.4 MeV (BBN energy region) has a contamination from the second 2^+ excited state of ^{8}Be. We could not separate the reaction which made the second 2^+ excited state of ^{8}Be with the high energy of deuteron ($\approx$2 MeV), since the incident energy was determined by the detected proton energy. However, this problem will be solved with using a lower incident deuteron

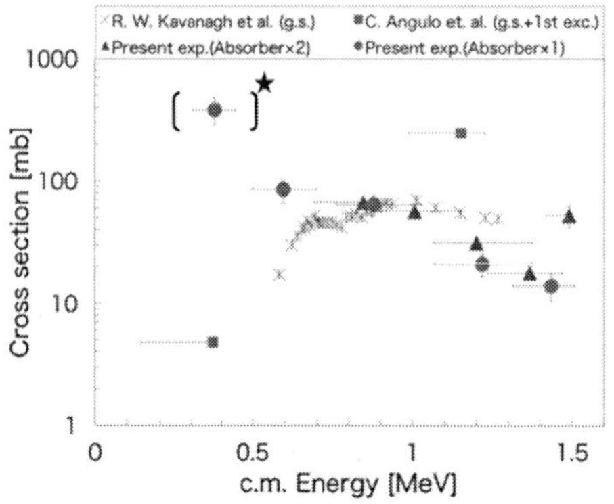

Figure 2. The cross section of the ^{7}Be(d, p) reaction as a function of the incident deuteron energy with the reference data [**? ?**]. This data was obtained using the activated ^{7}Be target. ⋆ This plot is preliminary.

energy that inhibits the higher excited state of ^{8}Be. An additional experiment is already planned at the Tandem Electrostatic Accelerator, Kobe University for March 2018. We plan to apply the method to produce ^{7}Be and measure the ^{7}Be(d, p)^{8}Be reaction.

2.2 Implantation method

The development of the implantation method is also ongoing. In the implantation method, ^{7}Be particles are implanted into a host target. We performed an experiment to produce a ^{7}Be implanted target at CRIB, Center for Nuclear Study (CNS), University of Tokyo, in June 2016. The primary beam was ^{7}Li^{2+} at 5.6 MeV/nucleon. The secondary beam was produced by the ^{1}H(^{7}Li, ^{7}Be) reaction. The secondary beam energy was 4.0 MeV/nucleon. A 10 μm thick gold foil as a host target was irradiated with the ^{7}Be beam after an energy degrader made of gold with a thickness of 15 μm and 2 mm diameter collimator. We evaluated the amount of the implanted ^{7}Be by detecting 477 keV γ rays with a $LaBr_3$ detector after the implantation. The γ ray is emitted in the electron capture decay of ^{7}Be with a branching ratio of 10.4%. We obtained 1.3×10^{11} ^{7}Be particles after 19 hours of irradiation.

We also performed a development experiment at CRIB to obtain an intense beam of ^{7}Be in 2017. We obtained 1.2×10^{12} ^{7}Be particles after the optimization of the magnets setup and settings of beam-line optics. As the next step, we plan to measure the ^{7}Be(d, p) reaction at Japan Atomic Energy Agency, tandem facility. The ^{7}Be target will be produced at CRIB before the (d, p) reaction measurement. The experiment is planned for 2018.

Acknowledgment

This work was supported by JSPS KAKENHI Grant Number JP 16H03980.

References

[1] R. H. Cyburt *et al.*, J. Cosmol. Astropart. Phys. **11**, 012 (2008).
[2] M. Barbagallo *et al.*, Phys. Rev. Lett. **117**, 152701 (2016)
[3] S. Q. Hou *et al.*, Phys. Rev. C **91**, 055802 (2015).

[4] T. Kawabata *et al.*, Phys. Rev. Lett. **118**, 052701 (2017).
[5] R. W. Kavanagh *et al.*, Nucl. Phys. **18**, 493 (1960).
[6] C. Angulo *et al.*, Astrophys. J. **630**, L 105 (2005).

A Geant4-based Monte Carlo Tool for Nuclear Astrophysics

D. Lattuada[1,*], *M.* La Cognata[2], *A.* Anzalone[2], *D.L.* Balabanski[1], *S.* Chesnevskaya[1], *M.* Costa[2], *V.* Crucillà[2], *G.L.* Guardo[1], *M.* Gulino[2,3], *C.* Matei[1], *R.G.* Pizzone[2], *S.* Romano[2,4], *C.* Spitaleri[2], *A.* Tumino[2,3], and *Y.* Xu[1].

[1]*ELI-NP / IFIN-HH, Magurele, Romania.*
[2]*INFN, Laboratori Nazionali del Sud, Catania, Italy.*
[3]*Facoltà di Ingegneria e Architettura, Università Kore, Enna, Italy.*
[4]*Dipartimento di Fisica e Astronomia, Università degli Studi, Catania, Italy.*

Abstract. Present and future gamma-beam facilities represent a great opportunity to validate and evaluate the cross-sections of many photonuclear reactions at near-threshold energies, whose data mostly come from theoretical calculations. We developed a Monte Carlo (MC) software that makes use of the validated tracking Geant4 libraries and the n-body event generator of ROOT libraries in order to provide a fast, realiable and complete MC tool to be used for nuclear physics experiments, with a particular focus on photo-nuclear processes. We discuss the results of the MC simulations performed in order to evaluate the effects of the electromagnetic background, the straggling of the emitted particles due to the target thickness and the resolution of the silicon detectors. Finally we present the preliminary results on some nuclear reactions involved in the p-process, which will be studied with ELISSA and the GBS at ELI-NP.

1 Introduction

Monte Carlo (MC) simulations have always played a crucial role in preparation of experiments, in data analysis and for proper understanding of the experimental results. With the development of new experimental methods and techniques, as well as more sensitive and complex detectors, we experience a growing need for details in the simulations. In this scenario the lack of a complete, validated, fast and open-source software tool to be jointly (commonly) used in nuclear physics experiments reflects the complexity of comparing data coming from different experiments. Therefore we are developing such a tool, using open-source scientific libraries with a strong record of validation in various fields of physics. This tool is intended to be used for studying photonuclear reactions at γ-beam facilities with ELISSA (Extreme Light Infrastructre Silicon Strip Array [1, 2]), a new silicon-strip detector array under development at the Extreme Light Infrastructure - Nuclear Physics (ELI-NP [3]) laboratory, but any detector and any nuclear reaction can be used. The idea of this toolkit came from the current lack of validated photonuclear *physics* working at low energies in Geant4 and the need of a Geant4-based code to provide detailed information on electromagnetic (and hadronic)

*e-mail: dario.lattuada@eli-np.ro

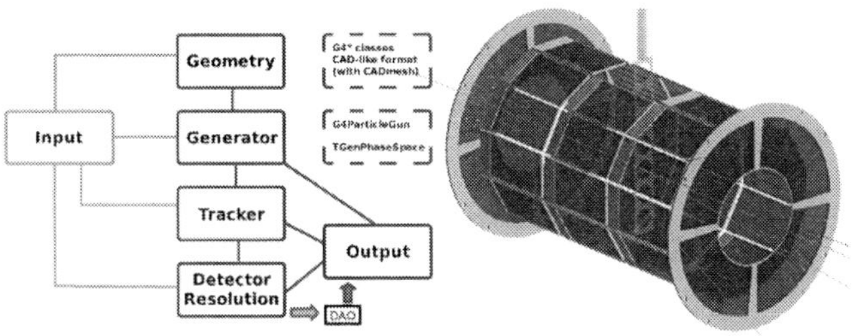

Figure 1. *Left*: A scheme of the MC code. *Right*: The geometry of ELISSA imported in GROOT. Gamma (*green*) and electron (*red*) tracks are shown.

background in nuclear physics experiments. The open-source GROOT [4] (Geant4 [5] and ROOT [6] Object-Oriented Toolkit) code we are currenlty developing mainly relies on ROOT, Geant4 and the Standard Template Library (STL) of C++.

2 Results

The tool has been and will be used for past and future experiments. With ELISSA at ELI-NP, it will be possible to study photonuclear reactions of great importance for astrophyisics. Here, we present some preliminary results obtained with GROOT involving the p-process, i.e. the mechanism thought to be responsible for the abundances of the so-called p-nuclei, stable proton-rich nuclides far from the stability line, typically less abundant by factors of ten to one thousand than the other isotopes of the same element. The used photonuclear reaction rates of (γ, α) and (γ, p) reactions are computed in [7] with the help of TALYS[8]. The choice of the model is to be considered as an example, since the generated events are weighted *a posteriori*. In this way, it is very easy to compare different model predictions for the same reaction (and we did it using other approaches in the past [4]), by simply renormalizing the reaction rates according to the evaluated cross-sections. We examine the ^{92}Mo$(\gamma, \alpha)^{88}$Zr and ^{92}Mo$(\gamma, p)^{91}$Nb photonuclear reactions at 9 MeV, of great interest for the p-process. The latter is very difficult to study in the inverse (p, γ) reaction, because of the very limited availability of ^{91}Nb targets and beams.

In figure 2, the recorded energy of the ejected particles is plotted (ELISSA setup is shown in figure 1 for reference, only the X3 "rings" are considered). The data are plotted after applying a Gaussian smearing with an energy-dependent sigma in order to simulate the detector energy resolution as reported from Micron and calculated in [9, 10]. Despite the straggling in the target, the expected alpha-signal is well separated from the proton one, which in turn is (calculated to be) much greater, by nearly 9 orders of magnitude. Theta vs. alpha spectra are used to disentangle the proton and alpha signals owing to the different kinematics, similarly to how we did in [4]. The electromagnetic background is calculated for a setup with a 25 μm thick aluminum flange foil placed 3 m far from a thin 3 μm ^{92}Mo target on a 0.5 μm carbon backing (black squares on figure 2, improvements on statistics are ongoing). By using an additional turbopump we can have 3 meters of beam line and the chamber under the same vacuum conditions, so that the photons will only encounter one thin window between the γ transport line and the scattering chamber at pressures of 10^{-3} mbar e 10^{-6} mbar respectively.

In conclusion, the validation of the tool is still in progress and we are currently analizing our experimental results from various tests and experiments, in order to compare them with

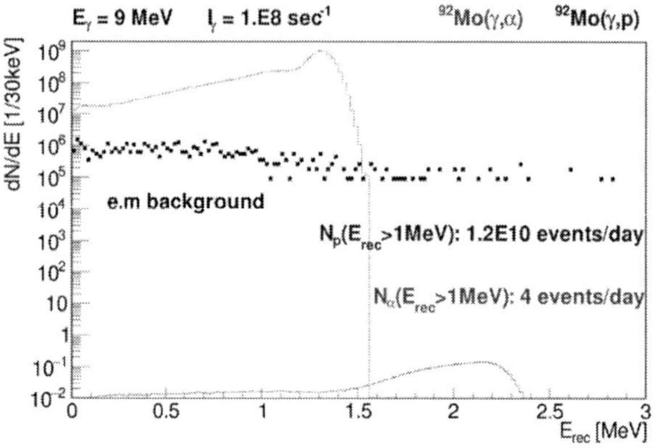

Figure 2. Alphas (*red* line) and protons (*blue* dotted line) coming from the γ+^{92}Mo reaction at 9 MeV beam with ELI-NP GBS specification. The black squares represent the evaluated e.m. background. I_γ is a conservatory estimate of the beam intensity, the number of generated background events thus roughly corresponds to 1 second livetime. further simulations of the e.m. background are going in order to increase the statistics

the simulations. Data from experiments are crucial in order to validate our tool and we look forward for future collaborations with international groups. Nevertheless, the reasonable results we show here are realiable since they are obtained using widely validated libraries and classes from Geant4 and ROOT. The code is currently under development at ELI-NP and due to its high level of customizability and ease to use it is applying to become the reference MC tool for photonuclear experiments with silicon detectors at this facility.

We acknowledge the support from the Extreme Light Infrastructure Nuclear Physics (ELI-NP) Phase II, a project cofinanced by the Romanian Government and the European Union through the European Regional Development Fund - the Competitiveness Operational Programme (1/07.07.2016, COP, ID 1334).

References

[1] O. Tesileanu et al., 2016 Charged particle detection at ELI-NP, Rom. Rep. Phys. 68 S699
[2] M. La Cognata et al., 2017 Journal of Instrumentation 12 C03079
[3] D. Filipescu et al., Eur. Phys. J., A51, 185 (2015)
[4] D. Lattuada et al., EPJ Web of Conferences 165, 01034 (2017)
[5] Geant4: A Simulation toolkit - Geant4 Collaboration (Agostinelli, S. et al.), Nucl. Inst. & Meth. A506 (2003) 250-303 SLAC-PUB-9350, FERMILAB-PUB-03-339.
[6] ROOT - An Object Oriented Data Analysis Framework - Rene Brun and Fons Rademakers, Proceedings AIHENP'96 Workshop, Lausanne, Sep. 1996, NIM-A 389 (1997) 81-86.
[7] Y. Xu et al., EPJ Web of Conferences 146, 01015 (2017)
[8] A.J. Koning, S. Hilaire, M.C. Duijvestijn, "TALYS-1.0", in Proceedings of the International Conference on Nuclear Data Science Technology 2007, EDP Sciences Nice, France (2008), p. 211-214.
[9] G. L. Guardo et al., EPJ Web of Conferences 165, 01026 (2017)
[10] S. Chesnevskaya et al., EPJ Web of Conferences 165, 01011 (2017)

Big Bang Nucleosynthesis (BBN) and Non-Standard Physics

Tahani Makki[1,*] and *Mounib* El Eid[1]
[1] *Department of Physics, American University of Beirut, Lebanon*

Abstract. A brief overview on standard big bang nucleosynthesis (shortly, SBBN) is presented. First, we describe the outcome of the SBBN concerning the abundances of the light elements up to ^7Li. A comparison with observations reveals a Lithium overproduction, which is not understood yet and is termed as "Cosmological Lithium Problem". Resolving that problem is not easy, since many aspects are involved which nuclear, astrophysical and even a non-standard scenario may be invoked. These items are described in some details owing to the limited available space.

1 Introduction

The standard big bang nucleosynthesis (SBBN) is well understood in the context of the standard model of particle physics. It predicts that the universe is composed of about 75% of Hydrogen and 25% of ^4He and small amounts of D, ^3He, ^7Li and ^6Li [4]. The most important nuclear reactions intervening in SBBN are given in Fig.1, while a comparison between SBBN calculations and observations is given in Table 1. It is clear that all abundances agree with observations except for Lithium which is higher by a factor of 3 than observations. This discrepancy constitute now one of the most intriguing problems in cosmology. In addition, Lithium was shown to have a constant behavior as function of metallicity, so called "Spite Plateau"[17]. Such a behavior is now in conflict with recent observations at very low metallicity which makes the Lithium problem more complicated. Solving the Lithium problem by taking into considerations accurate measurements of nuclear cross sections seems not possible [4]. In addition, the possibility of an astrophysical solution is also unlikely [13] because it will be constrained by the plateau at high metallicity.

2 Non-Standard Big Bang Nucleosynthesis

Including new particles/physics and changing fundamental constants were extensively studied to see the effect of these scenarios on SBBN or to constrain them by SBBN predictions. In addition, some non-standard scenarios were included for the aim to decrease the Lithium abundance. We mention a promising attempt by [7] where modifying the velocity distributions of nuclei during BBN have led to a decrease in Lithium to match observations.
In the following we give an example of non-standard scenario by adding extra dimension like dark component [1] and introducing a variable number of neutrinos and their chemical potential (β). Increasing the number of neutrinos and/or adding β will increase the standard radiation density ρ_{rad}

*e-mail: trm03@mail.aub.edu

1. $n \rightarrow {}^1H + e^- + \bar{\nu}$	7. ${}^3H + {}^4He \rightarrow {}^7Li + \gamma$
2. ${}^1H + n \rightarrow {}^2H + \gamma$	8. ${}^3He + n \rightarrow {}^3H + {}^1H$
3. ${}^2H + {}^1H \rightarrow {}^3He + \gamma$	9. ${}^3He + {}^2H \rightarrow {}^4He + {}^1H$
4. ${}^2H + {}^2H \rightarrow {}^3He + n$	10. ${}^3He + {}^4He \rightarrow {}^7Be + \gamma$
5. ${}^2H + {}^2H \rightarrow {}^3H + {}^1H$	11. ${}^7Li + {}^1H \rightarrow {}^4He + {}^4He$
6. ${}^2H + {}^3H \rightarrow {}^4He + n$	12. ${}^7Be + n \rightarrow {}^7Li + {}^1H$

Figure 1. The most important nuclear reactions for big bang nucleosynthesis.

Table 1. Comparison between SBBN predictions and observations.

	Predictions	Observations
Y_p	0.2457	0.2449 ± 0.0040 [2]
$D/H \times 10^5$	2.6719	$2.8^{+0.8}_{-0.6}$ [10]
${}^3He/H \times 10^5$	1.0193	1.1 ± 0.2 [3]
${}^7Li/H \times 10^{10}$	4.3003	$1.58^{+0.35}_{-0.28}$ [14]

shown in Eq.1. A second effect of β is altering the neutron to proton ratio at the freeze out temperature and this effect is limited to the electron neutrino owing to the involved weak interaction rates. The dark component will modify the expansion rate by additional effective dark fluid ρ_D (Eq.1).The expansion rate is related to the Hubble constant (H) via a modified Friedmann equation [1]:

$$\left(\frac{\dot{a}}{a}\right)^2 = H^2 = \frac{8\pi G}{3}(\rho_{rad} + \rho_D) \tag{1}$$

with ρ_D given by:

$$\rho_D(T) = k_\rho \rho_{rad}(T_0)\left(\frac{T}{T_0}\right)^{n_\rho} \tag{2}$$

where T_0 is chosen to be 1 Mev=1.16×10^{10}K , k_ρ is the ratio of effective dark fluid density over the total radiation density at T_0 and n_ρ characterizes the adiabatic expansion of the fluid. Modifying the energy density of the universe which is radiation dominated will alter the standard-time temperature relation, $t(s) = 1.39 \times \frac{k^{-0.5}}{(T(Mev))^2}$ [9], and consequently light element abundances. Note that k is the effective number of relativistic degrees of freedom due to photon, neutrino, electron and positron degrees of freedom. In Table 2 (second and third columns) we present the effect of dark component along with neutrino chemical potentials β (assumed to be equal for all species) where the calculated abundances agrees with observations. This is to confirm that SBBN is a rich ground for many non-standard physics.

On the other hand, it could happen that non-equality between chemical potential still exist. For example, hypothetical neutrino-majoron coupling can reopen the window of allowing large muon/tau neutrino asymmetry together with a small electronic asymmetry ([5]). In addition, one can assume one neutrino species mixes with a sterile neutrino resulting in chaotic amplification of electron neutrino chemical potential ([15]) or assuming just a non-vanishing electron neutrino chemical potential ([16]) in a Linearly Coasting Cosmology. In a universe preferring large lepton asymmetry([6], [8]), it is important to allow neutrino chemical potential to be different in the aim to understand the SBBN

Table 2. Effect of including non-standard neutrino properties and dark component.

	$n_\rho = 6.3, k_\rho = 0.1$		$n_\rho = 0, k_\rho = 0$
	$N_\nu = 3, \beta = 0.2$	$N_\nu = 4, \beta = 0.25$	$\beta_{\nu_e} = 0.23$, $\beta_{\nu_{\mu,\tau}} = 2$
Y_p	0.2515	0.2471	0.2319
$D/H \times 10^5$	2.7029	2.9517	3.6530
$^3He/H \times 10^5$	1.0243	1.0569	1.1430
$^7Li/H \times 10^{10}$	4.3602	3.8121	2.6550

and to investigate their effect on Lithium abundance. In this way we obtain constraints on these chemical potentials shown in Eq. 3 which lead to significant reduction of Lithium as shown in the fourth column of Table 2.

$$0.18 \leq \beta_{\nu_e} \leq 0.3, \quad 2 \leq \beta_{\nu_{\mu,\tau}} \leq 2.3 \tag{3}$$

As in most successful attempts to resolve the Lithium problem beyond the SBBN, in the present work the decrease in Lithium abundance is achieved at the expenses of increasing Deuterium. A remaining open scheme that is worth to investigate is the possibility of reducing Lithium abundance without much increase in Deuterium by including for example new particles/interactions. More detailed analysis of these non-standard physics will be presented in an upcoming paper by [18].

Acknowledgement

Tahani Makki would thank the organizing committee of the Ninth European Summer School on Experimental Nuclear Astrophysics for their support , especially she would thank prof. Rosario Gianluca Pizzone and Prof. Livio Lamia for their tremendous help.

References

[1] Arbey A. and Mahmoudi F., Phys. Lett. B 669, 46-51 (2008).
[2] Aver E., Olive K.A., and Skillman E.D., arXiv: 1503:08146 (2015).
[3] Bania T.M. et al., Nature 415, 54 (2002).
[4] Bertulani C.A. et al., arXiv:1802.03469v1, (2018).
[5] Dolgov A.D. and Takahshi F., Nucl. Phys. B, 688, 189-213 (2004).
[6] Hansen S. et al. , Phys. Rev. D 65, 023511 (2001).
[7] Hou S.Q. et al., ApJ 834, 165 (2017).
[8] Lunardini C. and Smirnov A.Yu., Phys. Rev. D 64, 073006 (2001).
[9] Mukhanov V., Physical Foundations of Cosmology (Cambridge University Press, New York, 2005) 421.
[10] Noterdaeme P. et al., A & A 542, L33 (2012).
[11] P.A.R. Ade et al., A & A 594, A13 (2016).
[12] Pizzone R.G., et al., ApJ 786, 112 (2014).
[13] Prantzos N. , A&A 542, A67 (2012).
[14] Sbordone L. et al., A&A 522, A26 (2010).
[15] Shi X., Phys. Rev. D 54, 2753 (1996).
[16] Singh P., JCAP 05, 061 (2015).
[17] Spite M. and Spite F., Nature 297, 483 (1982).
[18] T.R. Makki, M.F. El Eid, and G.J. Mathews, submitted to MNRAS (2018).

Study of key resonances in the ^{30}P$(p, \gamma)^{31}$S reaction in classical novae

A. Meyer[1,★], *N.* de Séréville[1], *F.* Hammache[1], *P.* Adsley[1], *M.* Assié[1], *D.* Beaumel[1], *C.* Delafosse[1], *F.* Flavigny[1], *A.* Georgiadou[1], *A.* Gottardo[1], *L.* Grassi[1], *J.* Guillot[1], *T.* Id Barkach[1], *M.* MacCormick[1], *I.* Matea[1], *L.* Olivier[1], *L.* Perrot[1], *C.* Portail[1], *I.* Stefan[1], *A.* Parikh[2], *A.* Coc[3], *J.* Kiener[3], *V.* Tatischeff[3], *A. M.* Laird[4], *S. P.* Fox[4], *N.* Hubbard[4], *J.* Riley[4], *F.* De Oliveira[5], *B.* Bastin[5], *K.* Béroff[6], *Á. M.* Sánchez Benítez[5,7], *A.* Alellara[8], *M.* Assunção[8], *V.* Guimaraes[8], *N.* Oulebsir[9], and *G.* D'Agata[10,11]

[1] *Institut de Physique Nucléaire, IN2P3/CNRS-Université de Paris XI, 91406 Orsay Cedex, France*
[2] *Departament de Física i Enginyeria Nuclear, Universitat Politècnica de Catalunya, E-08036 Barcelona, Spain*
[3] *Centre des Sciences Nucléaires et des Sciences de la Matière (CSNSM), IN2P3/CNRS-Université de Paris XI, 91405 Campus Orsay, France*
[4] *Department of Physics, The University of York, York YO10 5DD, United Kingdom*
[5] *Grand Accélérateur National d'Ions Lourds (GANIL), CEA/DRFCNRS/IN2P3, Bvd. Henri Becquerel, 14076 Caen, France*
[6] *Institut des Sciences Moléculaires d'Orsay (ISMO), UMR 8214, CNRS, Université Paris Sud*
[7] *Departamento de Ciencias Integradas, University of Huelva, 21071 Huelva, Spain*
[8] *Instituto de Física, Universidade de São Paulo, Rua do Matão 1371, 05508-090 São Paulo, Brazil*
[9] *Laboratoire de Physique Théorique, Université Abderahmane Mira, 06000 Béjaia, Algeria*
[10] *INFN-Laboratori Nazionali del Sud, Via S. Sofia, I-95125 Catania, Italy*
[11] *Dip. di Fisica e Astronomia, Università di Catania, Via S. Sofia, I-95125 Catania, Italy*

Abstract. Among reactions with strong impact on classical novae model predictions, ^{30}P$(p, \gamma)^{31}$S is one of the few remained that are worthy to be measured accurately, because of their rate uncertainty, as like as ^{18}F$(p, \alpha)^{15}$O and ^{25}Al$(p, \gamma)^{26}$Si. To reduce the nuclear uncertainties associated to this reaction, we performed an experiment at ALTO facility of Orsay using the ^{31}P$(^3$He$, t)^{31}$S reaction to populate ^{31}S excited states of astrophysical interest and detect in coincidence the protons coming from the decay of the populated states in order to extract the proton branching ratios. After a presentation of the astrophysical context of this work, the current situation of the ^{30}P$(p, \gamma)^{31}$S reaction rate will be discussed. Then the experiment set-up of this work and the analysis of the single events will be presented.

1 Introduction

Classical novae are stellar explosions occurring in binary systems consisting of a white dwarf accreting hydrogen-rich material from a companion main sequence star. This accreted matter is mixed

★e-mail: meyer@ipno.in2p3.fr

with the outer layers of the underlying white dwarf. This mixture is heated and compressed until a thermonuclear runaway takes place, leading to new elements ejected into the circumstellar medium. The overall characteristics of classical novae are well described by theoretical models but several key issues remain unexplained: the mixing mechanism and the degree of mixing between the outer white dwarf layers and the accreted material and the observed ejecta masses which are underpredict by an order of magnitude or more.

The study of nuclear reactions during the explosion allow to constrain the physical properties of classical novae. Hydrodynamic models [1] and post-processing calculations [2] have shown that the $^{30}P(p,\gamma)^{31}S$ reaction is a bottleneck for nucleosynthesis of nuclei up to Ca. Its rate therefore affects strongly the abundance predictions for these nuclei. It is also particularly important for understanding the large $^{30}Si/^{28}Si$ isotopic ratios found in presolar meteoritic grains, that are used to be classified as originating from a nova event [3]. Moreover, it was found that the $^{30}P(p,\gamma)^{31}S$ reaction rate uncertainty has the largest impact on the predicted ratios of Si/H, O/S, S/Al, O/P and P/Al, which can be used to constrain the degree of mixing [4] and the peak temperature [5] during the explosion.

2 Current situation and goal

The direct measurement of the $^{30}P(p,\gamma)^{31}S$ reaction rate is currently not feasible due to the difficulty to produce high intensity radioactive ^{30}P beam. So far, indirect methods were used to populate the states of the compound nucleus ^{31}S in the Gamow window, which corresponds, for temperature achieved in novae between 0.1 and 0.4 GK, to excitation energies up to 600 keV above the proton threshold (S_p = 6.131 MeV). More recent experimental works are focused on determining the energies, spins and parities of the states of astrophysical interest [6] [7] [8] [9] [10], leading to a more consistent picture of the level scheme of ^{31}S.

In order to calculate the reaction rate, the resonance strengths $(\omega\gamma)$ are needed:

$$(\omega\gamma) = \frac{(2J_R+1)}{(2J_p+1)(2J_{^{30}P}+1)}\frac{\Gamma_p\Gamma_\gamma}{\Gamma} \quad (1)$$

where J_R, J_p and $J_{^{30}P}$ are the spins of the resonance in ^{31}S, the proton and ^{30}P (ground state), respectively, and the total width of the resonance Γ is the sum of the proton and the γ-ray partial widths (Γ_p and Γ_γ respectively). Experimental constraints on strengths of key low energy resonances have recently been reported [10], and only one experiment so far has measured proton branching ratios for states above 6.7 MeV [11]. We performed a new measurement of proton branching ratios to extend the existing data to states at lower excitation energy and to reduce the current uncertainties using the $^{31}P(^3He,t)^{31}S^*(p)^{30}P$ reaction.

3 Experiment

We measured the $^{31}P(^3He,t)^{31}S^*(p)^{30}P$ reaction at ALTO facility of Orsay. A 25 MeV ^3He beam impinged onto a 60 $\mu g/cm^2$-thick ^{31}P target with a 100 $\mu g/cm^2$-thick carbon backing (see Fig. 1). The tritons were momentum analyzed with an Enge split-pole magnetic spectrometer [12] at 10° in the laboratory. They were detected and identified at its focal-plane by a position-sensitive gas chamber, a ΔE proportional gas-counter and a plastic scintillator. The decaying protons emitted in coincidence with the tritons were detected in 6 Double Sided Silicon Strip Detectors (DSSSDs) having 16 strips on each side and covering about 15 % of the solid angle. In this experiment we lowered the discriminator threshold in order to detect low energy protons associated to resonances of interest. Furthermore, given that the split-pole angle is 10° in the laboratory, this corresponds to $\theta_{c.m.} \simeq 28°$ for the recoiling

^{31}S nuclei, leading to c.m. angular coverage of the DSSSD array between [75.1°; 177.5°], thereby extending the angular range of Wrede *et al.* set-up [11].

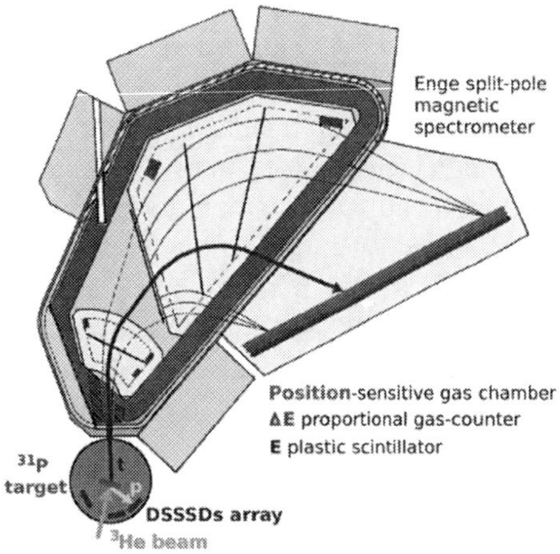

Figure 1. Schematic of the experimental set-up

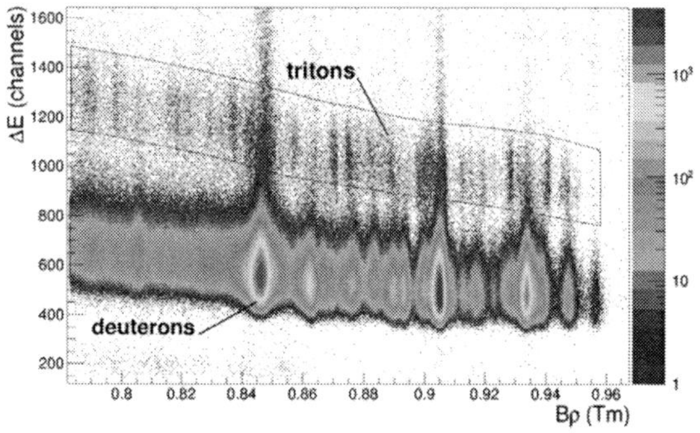

Figure 2. Bρ-ΔE identification of the light particles using the split-pole focal-plane detectors. One can see the deuteron loci extending into the triton region.

4 ^{31}P(^{3}He, *t*)^{31}S triton-singles spectrum

The calibration in magnetic rigidity (Bρ) of the focal-plane position detector was obtained using the triton spectrum at low excitation energies, taking into account the target and backing thickness. During the experiment the spectrometer magnetic field has been recorded, and showed unexpected variations in time, leading to Bρ variations. The maximum field variation observed was about 6×10^{-3} T, which corresponds to about 160 keV variation in excitation energy, while the energy resolution is 50 keV. In order to improve the spectrum quality the magnetic field and the position signals had to be precisely aligned in time.

Then the triton-singles spectrum at energies of interest was obtained after particle identification. However, as one can see in Fig. 2, the deuteron loci extend into the triton region. To remove this deuteron contamination, it was necessary to do several cuts using the signals recorded in the various focal-plane detectors, ie the position-sensitive gas chamber (Bρ), the proportional gas-counter (ΔE), the plastic scintillator (E_{PlasG} and E_{PlasP}) and the wire of the position-sensitive gas chamber (ΔE_{Fil}). The effect of the different cuts on the triton spectrum are shown in Fig. 3, considering only 7% of all statistic which has been analyzed so far. The comparison of the final spectrum, in red, with the level scheme observed by Wrede *et al.* [11], in blue, shows a satisfying agreement.

5 Perspectives

Analysis will proceed obtaining the coincidence spectrum and then extracting the proton branching ratios. This will be accomplished comparing the number of tritons detected in coincidence with

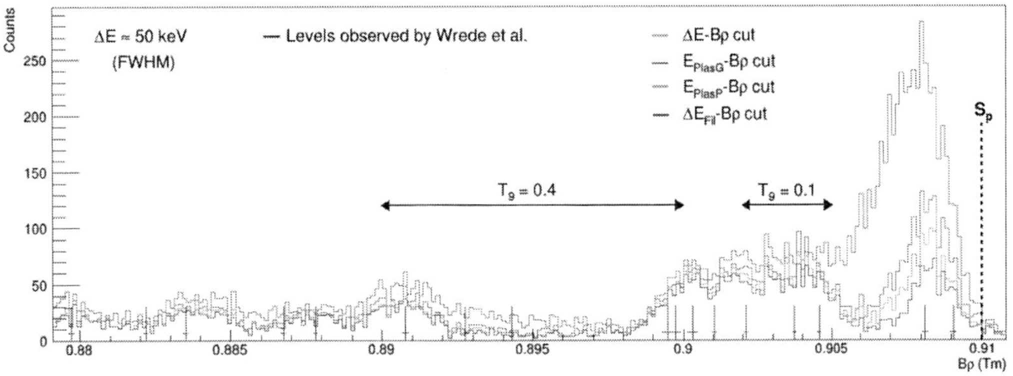

Figure 3. Focal-plane triton spectra from the ^{31}P(^{3}He, t)^{31}S reaction at 25 MeV and θ_{lab} = 10°, for 7% of all statistic. The effect of the different cuts using combination of focal-plane detectors signals are shown and the final spectrum (in red) can be compared to the levels observed by Wrede *et al.* (in blue).

decaying protons with the number of tritons detected without coincidence. Finally, the obtained proton branching ratios will be used in the ^{30}P(p, γ)^{31}S reaction rate calculation.

References

[1] J. José *et al.*, Astrophys. J. **560**, 897 (2001)
[2] C. Illiadis *et al.*, Astrophys. J. Supppl. **142**, 105 (2002)
[3] S. Amari *et al.*, Astrophys. J. **551**, 1065 (2001)
[4] K. J. Kelly *et al.*, Astrophys. J. **777**, 130 (2013)
[5] L. N. Downen *et al.*, Astrophys. J. **762**, 105 (2013)
[6] A. Parikh *et al.*, Phys. Rev. C **83**, 4, 045806 (2011)
[7] D. Irvine *et al.*, Phys. Rev. C **88**, 5, 055803 (2013)
[8] D. T. Doherty *et al.*, Phys. Rev. C **89**, 045804 (2014)
[9] B. A. Brown *et al.*, Phys. Rev. C **89**, 062801 (2014)
[10] A. Kankainen *et al.*, Phys. Lett. B **769**, 549 (2017)
[11] C. Wrede *et al.*, Phys. Rev. C **79**, 4, 045803 (2009)
[12] J.E Spencer and H. A. Enge, Nucl. Instrum. Methods **49**, 181 (1967)

Resonance Effects in Carbon Burning Process on Type Ia Supernovae

Kanji Mori[1,2,⋆], *Michael* Famiano[3,1,⋆⋆], and *Toshitaka* Kajino[1,2,4,⋆⋆⋆]

[1] *National Astronomical Observatory of Japan 2-21-1 Osawa, Mitaka, Tokyo, 181-8588 Japan*
[2] *Department of Astronomy, Graduate School of Science, The University of Tokyo, 7-3-1 Hongo, Bunkyo-ku, Tokyo, 113-0033 Japan*
[3] *Department of Physics, Western Michigan University, Kalamazoo, Michigan 49008 USA*
[4] *School of Physics and Nuclear Energy Engineering, Beihang University, Beijing 100191, China*

Abstract. The ^{12}C+^{12}C reaction plays an important role in type Ia supernovae. However, its cross section with astrophysical low energy has not been measured. If there is an unknown resonance in the low energy region, the reaction rate is enhanced and can affect astrophysics. We impose an upper limit to the resonance strength using the Wigner limit and find that the reaction rate can be larger than a standard rate by $\sim 10^3$ times in astrophysical temperature. We apply this enhanced reaction rate to binary mergers of two white dwarfs (WDs). We find that the fate of WD-WD mergers changes due to the resonance. The accretion induced collapse to neutron stars becomes more likely to occur, while it becomes more difficult to explain SNe Ia from WD-WD mergers.

1 Introduction

The C burning processes

$$^{12}\mathrm{C} +^{12}\mathrm{C} \rightarrow \alpha +^{20}\mathrm{Ne} + 4.6\ \mathrm{MeV} \quad (1)$$

$$^{12}\mathrm{C} +^{12}\mathrm{C} \rightarrow \mathrm{p} +^{23}\mathrm{Na} + 2.2\ \mathrm{MeV} \quad (2)$$

plays important roles in type Ia supernovae (SNe Ia) [2], X-ray superbursts [3] and massive stellar evolution [8]. Cross sections of this reactions have been pursued by a number of experiments, but cross sections with $E < 2.1$ MeV are still unknown (e.g. [4]). On the other hand, the Gamow peak at astrophysical temperature 0.5 GK is 1.5 MeV, therefore it is necessary to explore lower energy cross sections. A theoretical study [11] on the compound nucleus ^{24}Mg suggest that there are resonances around the Gamow peak. Therefore, in this study, we assume an unknown resonance, which will enhance the reaction rate and affect astrophysics.

Binary mergers of two white dwarfs (WDs) are thought to be a progenitor of SNe Ia [5, 6], while some suggest that they end up as neutron stars (NSs) instead [14]. The fate of WD-WD mergers is dependent on the ^{12}C +^{12} C reaction rate (e.g. [15]) because the C fusion ignites the explosion. The

⋆e-mail: kanji.mori@nao.ac.jp
⋆⋆e-mail: michael.famiano@wmich.edu
⋆⋆⋆e-mail: kajino@nao.ac.jp

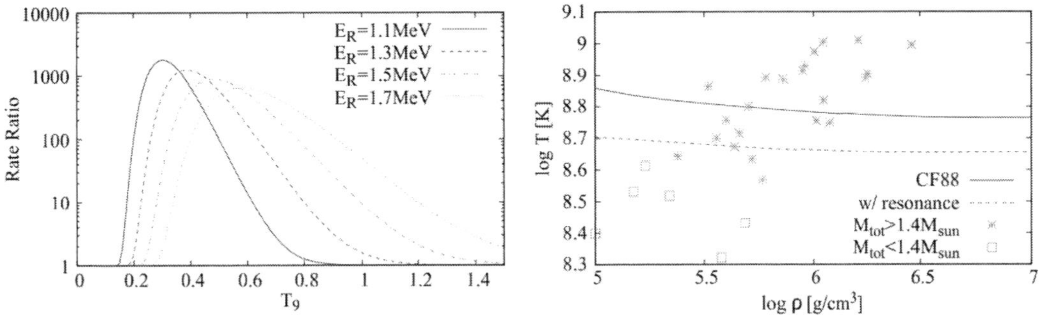

Figure 1. (Left) The upper limit of the reaction rate normalized by CF88. (Right) Ignition temperature of the $^{12}C+^{12}C$ reaction with and without the resonance. The points show the results of the SPH simulations [15].

aim of this study is to explore the effect of the low-energy resonance on the evolution of WD-WD mergers [1].

2 Constraint on the Resonance

The partial width of resonances must satisfy

$$\Gamma_C(E_R) < 2\gamma_W^2 P(E_R), \tag{3}$$

where E_R is the resonance energy, $\Gamma_C(E_R)$ is the partial width of the incoming channel, $P(E)$ is the penetration factor and $\gamma_W^2 = 3\hbar^2/(2\mu a^2)$ is the Wigner limit [7]. The penetration factor is given as

$$P(E) = \frac{ka}{G_l(ka)^2 + F_l(ka)^2}, \tag{4}$$

where k is the wave number and G_l and F_l are the Coulomb wave functions. The Wigner limit depends on the channel radius a. Kanungo et al. [17] measured the matter radius of ^{12}C as 2.35 fm. On the other hand, the channel radius is estimated to be $a = 7.97$ fm by fitting the cross section [18]. In order to estimate the upper limit of the strength, we adopt $a = 7.97$ fm in this study.

Assuming spin-0, the inequality (3) can be used as the upper limit on the strength of the low energy resonances. When $E_R = 1.5$ MeV, the upper limit on the resonance strength is 1.4×10^{-12} MeV. Hence it is shown that some of resonances adopted in previous work [2, 3, 8] are too strong in terms of the Wigner limit. It is also shown that known resonances listed in [9] satisfy the inequality (3). From this upper limit, the upper limit for the reaction rate is obtained. The left panel in Fig. 1 shows the upper limit for the reaction rate normalized by a standard rate [10] (hereafter CF88) as a function of $T_9 = T/(10^9$ K). One can see that the resonant reaction rate can be enhanced by $\sim 10^3$ times compared with CF88. Chiba & Kimura [11] suggested $E_R = 1.37$ MeV from their antisymmetrized molecular dynamics calculation on the compound nucleus ^{24}Mg. We adopt this resonance energy in the next section.

3 Implication to the Fate of WD-WD Mergers

The fate of WD-WD binary mergers depends on relations among dynamical timescale τ_{dyn}, heating timescale due to C burning $\tau_C = C_P T/\epsilon_C$ and cooling timescale due to neutrino emission $\tau_\nu = C_P T/\epsilon_\nu$.

C_P is specific heat, ϵ_C is the heating rate due to C burning and ϵ_ν is the cooling rate by thermal neutrinos [12]. Let us assume that the total mass of the two WDs is larger than the Chandrasekhar mass $M_{\rm ch}$. If a condition $\tau_C < \tau_{\rm dyn}$ is satisfied just after merging, the system explodes as a SN Ia. This scenario is called violent merger (VM; [13]). If this condition is not satisfied, the merger creates a remnant. In this phase, the system can be burnt into oxygen-neon-magnesium WD by C burning and collapse to a NS if $\tau_C < \tau_\nu$. This evolutionary path is called accretion induced collapse (AIC; [14]). If this does not occur, C burning begins at the center and the system explodes as a SN Ia. We call this path accretion induced explosion (AIE). If the total mass is smaller than $M_{\rm ch}$, a massive WD (MWD) remains.

The right panel of Fig. 1 shows ignition temperature defined as $\tau_C = \tau_\nu$. The points are temperature and density of the hottest points in the remnant phase in a smoothed particle hydrodynamics (SPH) simulation [15]. The resonance lowers ignition temperature, therefore more systems go to AIC, while SNe Ia become less likely to occur. WD mergers with the primary mass of $0.9M_\odot$ is led to AIE with the CF88 rate, but the resonance can change their fate into AIC.

The mass distribution of WDs is estimated from Sloan Digital Sky Survey (SDSS) [16]. Galactic WD-WD merger rate is estimated as 1.4×10^{-13} $/M_\odot$/yr from SDSS [19]. One can estimate astrophysical event rates from these observations and the theoretical consideration on the fate of WD mergers. The VM rate is estimated to be 8.4×10^{-17} $/M_\odot$/yr for both of CF88 and the resonant reaction rate. The AIE rate is 1.3×10^{-14} $/M_\odot$/yr for CF88 and 8.3×10^{-15} $/M_\odot$/yr for the resonant rate. On the other hand, Galactic SN Ia rate is estimated to be $\sim 1.1\times10^{-13}$ $/M_\odot$/yr [20]. Hence WD-WD mergers explain only $\sim$ 12% of SNe Ia with CF88 and $\sim$ 8% with the resonant rate.

The authors thank S. Kubono for helpful discussion. KM was supported by a grant from the Hayakawa Satio Fund awarded by the Astronomical Society of Japan.

References

[1] Mori, K. Master Thesis (The University of Tokyo); to be submitted to ApJ
[2] Bravo, E. et al. A&A **535**, A114 (2011)
[3] Cooper, R. L., Steiner, A. W., & Brown, E. F. ApJ **702**, 660 (2009)
[4] Spillane, T. et al. PRL **98**, 122501 (2007)
[5] Iben, I. Jr., & Tutukov, A. V. ApJS **54**, 335 (1984)
[6] Webbink, R. F. ApJ **277**, 355 (1984)
[7] Teichmann, T., & Wigner, E. P. PR **87**, 123 (1952)
[8] Bennett, M. E. et al. MNRAS **420**, 3047 (2012)
[9] Aguilera, E. F. et al. PRC **73**, 064601 (2006)
[10] Caughlan, G. R., & Fowler, W. A. ADNDT **40**, 283 (1988)
[11] Chiba, Y., & Kimura, M. PRC **91**, 061302 (2015)
[12] Itoh, N. et al. ApJS **102**, 411 (1996)
[13] Pakmor, R. et al. Nature **463**, 61 (2010)
[14] Nomoto, K., & Kondo, Y. ApJ, **367**, L19 (1991)
[15] Sato, Y. et al. ApJ **807**, 105 (2015)
[16] Kepler, S. O. et al. MNRAS **375**, 1315 (2007)
[17] Kanungo, R. et al. PRL **117**, 102501 (2016)
[18] Yakovlev, D. G. et al. PRC **82**, 044609 (2010)
[19] Badenes, C., & Maoz, D. ApJ **749**, L11 (2012)
[20] Li, W. et al. MNRAS **412**, 1473 (2011)

The Treiman-Yang Criterion: validating the Trojan Horse Method by experimentally probing the reaction mechanism

S. S. Perrotta[1,2,3,★], *C.* Spitaleri[1,2], *S.* Cherubini[1,2], *A.* Cvetinović[2], *G.* D'Agata[1,2], *D.* Dell'Aquila[4], *A.* Di Pietro[2], *P.* Figuera[2], *L.* Guardo[2], *M.* Gulino[2,5], *I.* Indelicato[2], *I.* Kres[2], *M.* La Cognata[2], *M.* La Commara[4,6], *L.* Lamia[2], *D.* Lattuada[2,7], *M.* Lattuada[1,2], *I.* Lombardo[4,6], *M.* Mazzocco[8], *T.* Parascandolo[4,6], *R. G.* Pizzone[2], *G. G.* Rapisarda[2], *S.* Romano[1,2], *R.* Spartà[2], *O.* Trippella[9], and *A.* Tumino[2,5]

[1] *Dipartimento di Fisica e Astronomia, Università degli studi di Catania*
[2] *Istituto Nazionale di Fisica Nucleare, Laboratori Nazionali del Sud, Catania*
[3] *Scuola Superiore di Catania*
[4] *Istituto Nazionale di Fisica Nucleare, Sezione di Napoli*
[5] *Facoltà di Ingegneria e Architettura, Università degli Studi di Enna "Kore"*
[6] *Dipartimento di Scienze Fisiche, Università "Federico II", Napoli*
[7] *Extreme Light Infrastructure – Nuclear Physics*
[8] *Dipartimento di Fisica e Astronomia, Università degli Studi di Padova and INFN, Sezione di Padova*
[9] *Dipartimento di Fisica e Geologia, Università degli Studi di Perugia and INFN, Sezione di Perugia*

Abstract. Proper selection of the quasi-free (QF) break-up channel in a three-body reaction is a key aspect for the applicability of the Trojan Horse Method (THM). The Treiman-Yang (TY) Criterion is a model-independent experimental test for the dominance of the QF mechanism, and hence constitutes one of the strongest validity tests of the THM. An experiment was performed at LNS to apply the test to the $d(^{10}B, {}^{7}Be\ \alpha)n$ reaction. Here, the criterion is described and some preliminary data from the experiment are shown.

1 Introduction

The quasi-free break-up mechanism, depicted in figure 1a for the $d + {}^{10}B \rightarrow {}^{7}Be + \alpha + n$ reaction, can be described as the virtual interaction (here $p + {}^{10}B \rightarrow {}^{7}Be + \alpha$) of the incident particle (here ^{10}B) with only a cluster (here p) of the other reactant, named 'participant', emitted in a virtual decay (here $d \rightarrow p + n$), while the other cluster remains undisturbed. Under suitable conditions, for instance if the participant is non-relativistic (small momentum transfer) and has spin 1/2 (see [1] for a more detailed discussion), the spin-averaged square-modulus amplitude $\overline{|M|^2}$ for the QF channel can be factorized into the amplitudes for the two virtual processes and the propagator.

The Trojan Horse Method (see [2] for a review and [3] for a THM measurement of the $p + {}^{10}B \rightarrow {}^{7}Be + \alpha$ via the $d(^{10}B, {}^{7}Be\,\alpha)n$) is an indirect technique for measuring cross sections of astrophysically relevant two-body reactions, by selecting the QF channel of an appropriate three-body reaction. In THM measurements, the most typical competitor to the QF is the sequential decay

★e-mail: perrotta@lns.infn.it

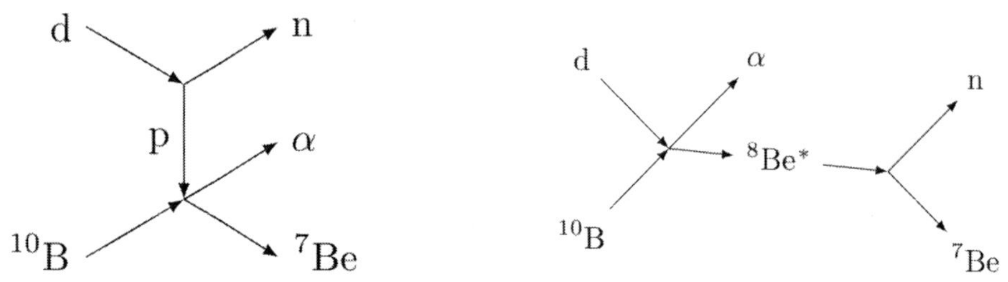

(a) Quasi-Free Breakup. (b) Sequential Decay.

Figure 1. Two of the infinitely many diagrams that could contribute to the $d + {}^{10}B \rightarrow {}^{7}Be + \alpha + n$.

(figure 1b), but in principle many other mechanisms can contribute to the reaction amplitude. It is thus pivotal to correctly select the kinematical locus where the QF channel dominates: several validity tests are usually employed in the THM analysis (see [2, 3]), but none of them is sufficient to completely ensure the dominance of the QF mechanism.

2 The Treiman-Yang Criterion

The Treiman-Yang Criterion was first proposed for high energy particle interactions [4] (where the QF was initially observed), then extended to non-relativistic nuclear reactions [1]. It is a very strong, model-independent experimental test for the preponderance of the quasi-free break-up mechanism.

In general, the spin-averaged amplitude for a three-body reaction depends on 5 kinematical variables. It is useful to choose them among the Mandelstam variables $s_{12} = |\mathbb{P}_1+\mathbb{P}_2|^2$ and $t_{12} = |\mathbb{P}_1-\mathbb{P}_2|^2$, where $\mathbb{P}_i$ is the four-momentum of particle i, or their non-relativistic analogues (defined in [1]): a convenient set for the $d + {}^{10}B \rightarrow {}^{7}Be + \alpha + n$ reaction is $s_{dB}, t_{dn}, t_{B\alpha}, s_{nBe}, s_{\alpha Be}$. The amplitude of any specific channel will depend on a subset of these variables.

It can be shown [1] that, if the amplitude $\overline{|M|^2}$ of the QF channel of a reaction is factorizable (as in section 1), then it will be a function of $t_{dn}, s_{\alpha Be}, t_{B\alpha}$ only. One can then experimentally check that $\overline{|M|^2}$ doesn't vary with s_{dB} and s_{nBe}, to rule out any mechanism whose amplitude depends on them.

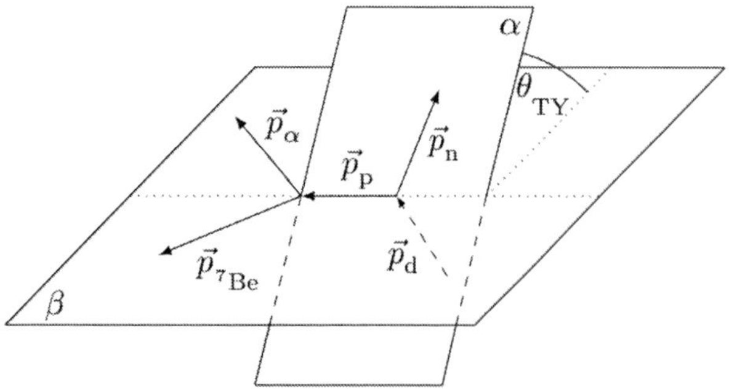

Figure 2. Representation (introduced by [5]) of the particles' momenta before and after the reaction $d + {}^{10}B \rightarrow \alpha + {}^{7}Be + n$ in the ${}^{10}B$ rest frame, given that it proceeds through QF breakup. θ_{TY} is the angle between the $(\vec{p}_d, \vec{p}_n)$ plane (α) and the $(\vec{p}_\alpha, \vec{p}_{Be})$ plane (β).

It is readily seen that, regardless of the reaction mechanism, $t_{dn}, s_{\alpha Be}, t_{B\alpha}$ and s_{dB} are all constant under a "Treiman-Yang rotation", defined in the $\vec{p}_B = 0$ frame as the rotation of $\vec{p}_\alpha$ and $\vec{p}_{Be}$ around $\vec{p}_\alpha + \vec{p}_{Be}$. Thus, if the reaction proceeds through QF breakup and factorization is possible, $\overline{|M|^2}$ must be constant under TY rotations. The particles' momenta then follow the scheme in figure 2: in the

^{10}B rest frame, $\vec{p}_p$ equals $\vec{p}_\alpha + \vec{p}_{Be}$ and it's the intersection between the planes $(\vec{p}_d, \vec{p}_n)$ and $(\vec{p}_\alpha, \vec{p}_{Be})$, so the TY rotation corresponds to rotating the two planes by an angle θ_{TY}.

A kinematical region of a reaction is therefore said to pass the TY Criterion when $\overline{|M|^2}$ doesn't depend on θ_{TY}. Although not a sufficient condition, this test is a strong signature for QF dominance.

3 The Experiment

Previous experimental attempts to apply the Treiman-Yang Criterion for reactions at low beam energies date back to 1980 [5]. In June 2016, a dedicated experiment was carried out at Laboratori Nazionali del Sud in Catania in order to verify the Criterion for the d(^{10}B, ^{7}Beα)n. The measurement took place in Camera2000, with a 27.5 MeV ^{10}B beam coming from the Tandem Accelerator and impinging on a 142 μg/cm^2 thick CD_2 target. Figure 3 shows a scheme of the setup.

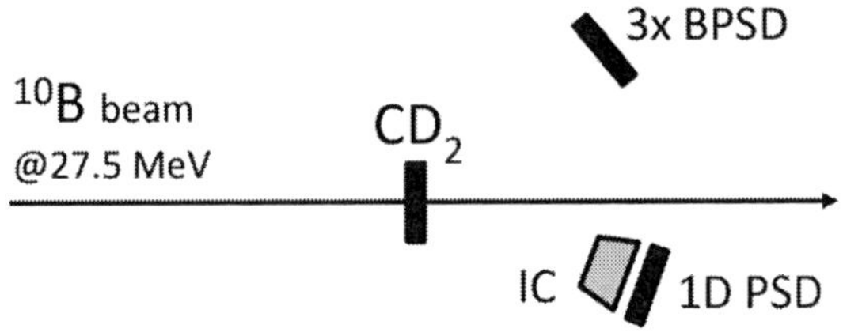

Figure 3. Schematic view of the experimental setup.

As is clearly seen in figure 2, in order to explore different θ_{TY} it's necessary to measure some reaction products out of the horizontal plane: a vertical stack of three Bidimensional Position Sensitive Detectors (BPSD) was therefore employed to detect the α particles. Beryllium ions were instead detected on-plane and identified by a ΔE-E telescope (see figure 4) made with an Ionization Chamber (IC) filled with isobutane gas placed in front of a one-dimensional Position Sensitive Detector (PSD).

The analysis of the data acquired in the experiment is still in progress. The detectors have been calibrated, and good energy and position resolution has been obtained for both the PSD and BPSD: for illustration, in figure 5 it is seen that the reference holes in the calibration grid of a BPSD are reproduced quite well. The results obtained from the partial and preliminary analysis make us confident that the experimental run was technically successful.

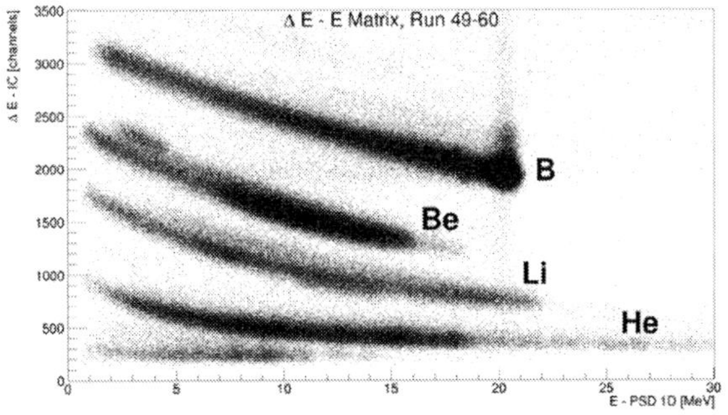

Figure 4. Typical ΔE-E spectrum from the IC-PSD telescope.

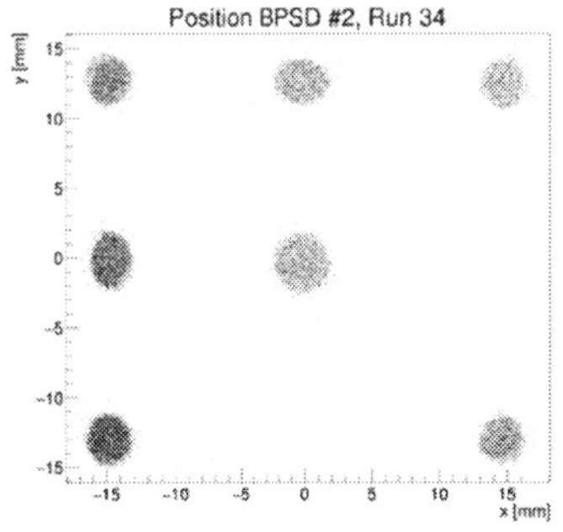

Figure 5. Measured position of particles that hit a BPSD preceded by a grid.

References

[1] I. S. Shapiro et al., Nuclear Physics **61**, 353–367 (1965).
[2] R. E. Tribble et al., Rep. Prog. in Phys. **77**, 106901 (2014).
[3] C. Spitaleri et al., Physical Review C **95**, 055806 (2017).
[4] S. Treiman and C. Yang, Phys. Rev. Lett. **8**, 140 (1962).
[5] P. G. Fallica et al., Physical Review C **24**, 4 (1981).

Isomeric ^{26}Al beam production with CRIB

H. Shimizu[1,⋆], *D.* Kahl[2], *H.* Yamaguchi[1], *K.* Abe[1], *O.* Beliuskina[1], *S. M.* Cha[3], *K. Y.* Chae[3], *A. A.* Chen[4], *Z.* Ge[5], *S.* Hayakawa[1], *N.* Imai[1], *N.* Iwasa[6], *A.* Kim[7], *D. H.* Kim[7], *M. J.* Kim[3], *S.* Kubono[5], *M. S.* Kawag[3], *J.* Liang[4], *J. Y.* Moon[8], *S.* Nishimura[5], *S.* Oka[9], *S. Y.* Park[7], *A.* Psaltis[4], *T.* Teranishi[9], *Y.* Ueno[9], and *L.* Yang[1]

[1] *Center for Nuclear Study, Department of Physics, the University of Tokyo, Japan*
[2] *School of Physics and Astronomy, the University of Edinburgh, United Kingdom*
[3] *Department of Physics, Sungkyunkwan University, Korea*
[4] *Department of Physics and Astronomy, McMaster University, Canada*
[5] *RIKEN Nishina Center, RIKEN, Japan*
[6] *Department of Physics, Tohoku University, Japan*
[7] *Department of Physics, Ewha Womans University, Korea*
[8] *Institute for Basic Science, Korea*
[9] *Department of Physics, Kyushu University, Japan*

Abstract. We performed an experiment to measure proton resonant elastic scattering of a mixed ^{26m,g}Al beam with a thick target in inverse kinematics by using CNS RI beam separator, located at RIKEN Nishina Center. It aimed to search for strong proton resonances and determine level properties of low spin-parity states in ^{27}Si. Diagnosis of the ^{26m}Al purity of the beam by annihilation radiation are discussed.

1 Introduction

Galactic abundance of the ^{26}Al radionuclide provides a unique window to the ongoing nucleosynthesis in the Milky Way. ^{26}Al is known as the first detected specific radioactivity that decays along with its characteristic β-delayed γ-ray and it has been directly observed by astronomical telescopes [1]. Despite a lot of effort over the past three decades, particular sites of galactic ^{26}Al are still poorly understood, and there is a discrepancy between observations and theories on estimations of its abundance.

However, this problem is complicated by its isomer, ^{26m}Al, which has a low spin $J^\pi = 0^+$ and a short lifetime $T_{1/2} = 6.35\ s$ compared with $T_{1/2} = 0.72$ Myr and $J^\pi = 5^+$ of ground state, and thus it directly decays to a stable state without emitting γ-ray. It is proposed that the ground state of ^{26}Al can communicate with the isomeric state through thermal excitations under high temperature environment such as core-collapse supernovae [2], therefore the reactions of the isomer might play an important role in the production of ^{26}Al in our Galaxy. Though the isomeric state of ^{26}Al might be a key to solve the discrepancy, only a little experimental information on the isomer has been published and theoretical calculations based on Hauser-Feshbach theory are used in the reaction networks so far. For

⋆e-mail: shimiz@cns.s.u-tokyo.ac.jp

this reason, the production of ^{26m}Al beam is a useful experimental approach for measuring nuclear properties and cross-sections as inputs to calculations of stellar reaction rates.

2 Production of ^{26m}Al

We performed an experiment with the Center for Nuclear Study low-energy RI Beam separator, CRIB, which is located at RIBF of the RIKEN Nishina center.

2.1 RI beam production with CRIB

A primary beam of $^{26}Mg^{8+}$ was supplied by an azimuthally varying field (AVF) cyclotron with radio frequency at 65 MHz and the kinetic energy of 172.9 MeV (6.65 MeV/u) and typical primary intensity of 20 – 50 pnA. A cryogenic hydrogen gas target system was used as a proton target to induce the $^{26}Mg(p,n)^{26}Al$ reaction in inverse kinematics. The secondary beam was bent by two dipole magnets, and by selecting the beams with a slit which is placed at the momentum-dispersive focal plane between two dipole magnets, the beam was purified according to the magnetic rigidity $B\rho$ and the slit width. The beam spot as well as the number of particles was measured by parallel plate avalanche counters at dispersive-focal plane after that. The isotopes in the beam were identified by energy versus time-

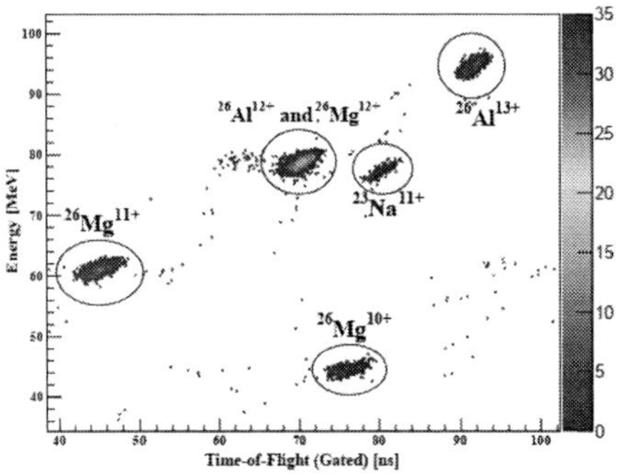

Figure 1. RI species identified at focal plane after the dipole magnets

of-flight plot, shown in Fig. 1. Through the Wien-Filter, $^{26}Al^{13+}$ was purified to more than 90% at the experimental target position, and typical intensity was around 1.5×10^5 pps, and typical energy was 114–120 MeV. However, because the ground state and the isomeric state of ^{26}Al have only a small mass difference (228 keV), they could not be distinguished event-by-event by the CRIB separator. By varying the secondary beam production target, H_2 gas pressure, several conditions of ^{26m}Al beam were produced during the experiment. Since CRIB uses a in-flight method and has a typical fight time of ^{26}Al on the order of 500 ns, the second excited state of ^{26}Al (417 keV) decays enroute to the traget, contributing to the yield of ^{26g}Al.

2.2 Determing purity of the isomers

To obtain the absolute purity of the isomers, the beam was in a pulsing mode with a cycle of 24 s for implantation, where the beam was on for 12 s for implantation and off for 12 s independent from scattering measurement. The RI beam was implanted into several targets such as CH_2 film and thick blocks, and when the beam was off, the radiation was measured by 10 NaI scintillators placed above the center of the target, in order to measure annihilation radiation from ^{26m}Al. The half life of the radiation at 511 ± 60 keV, where 60 is about 3σ of the energy resolution of the detectors, was measured to be 6.359 ± 0.150 s (cf. known $T_{1/2}$ of ^{26m}Al is 6.35 s). Therefore it is confirmed that almost all of the decay events originated from ^{26m}Al.

As ^{26}Al decays, positrons emitted from the inside target isotropically can escape to all the surrounding material and made a precise counting annihilation radiation difficult. To estimate such complicated situation, we defined the intrinsic efficiency η of the detectors as η = (511 keV counts)/(^{26m}Al decays) which depends on solid angle of detectors, quantum efficiency and geometry of the experimental chamber, and the Monte-Carlo simulation framework, Geant4, was used to evaluate it. By comparing simulation results and experimental data, the validity of the simulation was confirmed within an uncertainty of 5%.

2.2.1 Result of the decay measurement

Using the value of η from the simulation, the isomeric purities determined are tabulated in Tab. 1 under several conditions. These differences will be used to evaluate the contribution of ^{26g}Al to the cross section of the resonant scattering.

Table 1. Result of the isomeric purities under the different situations labeled as #1–#5, where #3 and #5, #2 and #4 are somewhat similar conditions respectively.

RI beam label	#1	#2	#3	#4	#5
Isomeric Purity (%)	48.3 ± 2.9	43.0 ± 2.5	56.4 ± 3.4	43.9 ± 2.4	52.7 ± 2.4

3 Conclusion

The isomeric RI beam ^{26m}Al was produced by CRIB with the purity of at maximum 56.4% and at minimum 43.0% according to the evaluation with the Monte-Carlo simulation. The energies of RI beams can cover the region of interest of high temperature environments such as core-collapse supernovae. With these isomeric ^{26}Al beams, proton elastic scattering was measured with a thick target in inverse kinematics method, in order to encompass the uncertainty of recent ^{26}Al problem connected with the observation of the γ-ray and the reaction networks of nuclear astrophysics.

References

[1] Diehl, Roland and Halloin, Hubert and Kretschmer, Karsten and Lichti, Giselher G and Schönfelder, Volker and Strong, Andrew W and Von Kienlin, Andreas and Wang, Wei and Jean, Pierre and Knödlseder, Jürgen and others, *Radioactive* ^{26}Al *and Massive Stars in the Galaxy*, arXiv preprint astro-ph **0601015** (2006)

[2] R.C.Runkle, A.E.Champagne and J.Engel, *Thermal Equilibration of* ^{26}Al, The Astrophysical Journal **446**:970–978 (2001)

[3] Yanagisawa, Y and Kubono, S and Teranishi, T and Ue, K and Michimasa, S and Notani, M and He, JJ and Ohshiro, Y and Shimoura, S and Watanabe, S and others, *Low-energy radioisotope beam separator CRIB*, Nuclear Instruments and Methods in Physics Research Section A: Accelerators, Spectrometers, Detectors and Associated Equipment **539(1)**:74–83 (2005)

^{26}Mg target for nuclear astrophysics measurements

I. Siváček[1,*], *J.* Mrázek[1], *V.* Kroha[1], *V.* Burjan[1], *V.* Glagolev[1], *Š.* Piskoř[1], *C.* Spitaleri[2], *R.G.* Pizzone[2], *M.* La Cognata[2], *G.* D'Agata[2], *A.* Tumino[2], *L.* Lamia[2], *R.* Spartá[2], *G.G.* Rapisarda[2], and *S.* Romano[2]

[1] *Nuclear Physics Institute of the Czech Academy of Sciences, p.r.i., Řež 130, 25068 Řež, Czech Republic*
[2] *Istituto Nazionale di Fisica Nucleare, Laboratori Nazionali del Sud, Via S. Sofia 62, 95125 Catania CT, Italy*

Abstract. Two nuclear reactions of astrophysical interest, ^{26}Mg(^{3}He,d)^{27}Al and ^{26}Mg(d,p)^{27}Mg, were measured for extraction of the Asymptotic Normalization Coefficients. Investigation of the target composition is presented, as well as the effects that showed up during analysis of the in-beam data obtained on CANAM accelerators in the Nuclear Physics Institute of the Czech Academy of Sciences (NPI CAS).

1 Introduction

One of main challenges of nuclear astrophysics is to explain origins and ratios of observed chemical elements abundances, since there are still questions in heavier elements production. Abundance of ^{27}Al is an essential check of models describing ^{26}Al production in Mg-Al cycle compared to observation of gammas from ^{26}Al decay in the galaxy [1]. Next challenge is to understand quantitatively the observed galactic distribution of ^{26}Al via ratio between ground and isomeric state in ^{26}Al in novae. This ratio may be influenced by destruction of ^{26}Si, which decays to ^{26}Al [2] . Nuclear reactions of charged particles in stars proceeds with energies below Coulomb barrier, so the cross sections are very low. Direct measurements in laboratory conditions are difficult and sometimes not feasible, therefore indirect methods are used. One of them is the method of "Asymptotic Normalization Coefficients" (ANC). This technique allows to extract normalization of nuclear overlap function tail for single particle transfer, which determines the direct capture reaction rate and the cross section of direct process. ANC can be evaluated from the analysis of direct nuclear reactions measured in laboratory conditions (elastic scattering, transfer reactions) [3, 5]. Using ^{26}Mg target, the studies of ^{26}Mg(d,p)^{27}Mg (as a mirror nucleus to ^{26}Si(p,g)^{27}P) and ^{26}Mg(^{3}He,d)^{27}Al reactions were measured to extract ANC.

2 Setup

Measurement of an angular distribution of reaction products is necessary for extraction of ANC. ^{26}Mg targets were used for in-beam irradiation of cyclotron U120M in NPI CAS. This cyclotron is operated under "Center of Accelerators and Nuclear Analytical Methods" (CANAM infrastructure). Angular distributions of the nuclear reaction products were measured with a set of dE/E telescopes consisting of 2 silicon detectors, 500 um and 5 mm thick respectively on VME ADC electronics. Beam energy

*e-mail: sivacek@ujf.cas.cz

of $^3He^{2+}$ was 25 MeV with typical intensities up to 20 enA. Set of data was acquired by 4 moving telescopes, covering angles from 7 to 61 degrees in laboratory system and 2 stationary telescopes as "beam monitors" (angles of 15 and 35 degrees was chosen for $^{26}Mg(^3He,d)$ reaction). For this type of measurement, thin target foils containing pure target isotope are preferred. Targets containing ^{26}Mg on carbon backing were used for the measurement, together with mylar target and pure carbon foil, with similar properties to magnesium target backing. Thin dE detectors were calibrated with $^{244}Cm+^{238}Pu$ alpha particle source. A number of peaks from ($^3He,^4He$) reactions, together with ($^3He,^3He$) scattering peaks on ^{12}C and ^{16}O nuclei, were identified in the dE detector energy spectrum. They were used for precise calibration of detectors with a simultaneous fine variations of the beam energy (T_{beam}) and detector angles. Thick E-detectors were calibrated on the basis of dE calibration and the deuterons from the reaction. Reaction products from $^{26}Mg(^3He,d)$ were identified in deuteron reaction channel afterwards. A complex program [5] was used for kinematic calibration, which allows to fit energy calibration of detectors, T_{beam} and detectors angles for all telescopes simultaneously. In the result, it allows to identify every peak in the energy spectrum and extract their angular distributions.

3 Target characteristics

Identification of peaks in the above-mentioned energy spectra allows to qualitatively characterize chemical composition of target. Target thickness was measured using energy loss method for alpha particles, measured with $^{244}Cm+^{238}Pu$ alpha particle source. Energy loss of alpha particles in the medium of known chemical composition is proportional to its thickness. Results obtained from energy loss tables for alpha particles were confirmed with Monte Carlo simulations in TRIM. Data from the "beam-monitor" telescope were used to determine stoichiometric proportions of the magnesium target and to check the beam energy together with stability during the irradiation. Stoichiometry of magnesium target was done by evaluation of ^{16}O reactions yields in mylar target, which was irradiated as well as carbon foil target. Yields of $^{16}O(^3He,^4He)^{15}O$ reaction are crucial, due to assumption, that oxygen in the target belongs to MgO. Analysis of spectra obtained with carbon target allowed to evaluate thickness of carbon backing in the magnesium target. It has uncovered an oxygen admixture in carbon backing, which was considered as the "background oxygen" and subtracted from oxygen yields in the magnesium target. Based on oxygen yields obtained with the mylar target, the amount of magnesium (corresponding to MgO) was evaluated to 135,1 +/- 6,7 $\mu g/cm^2$, as a mean value obtained from the reactions $^{16}O(^3He,^3He)^{16}O$ (highest yields on a peak background) and $^{16}O(^3He,^4He)^{15}O$ to the ground state (no peak background). This value was used to evaluate differential cross sections (quantitatively evaluate angular distributions) of $^{26}Mg(^3He,d)^{27}Al$.

4 Discussion

During the data analysis, a number of effects appeared with the impact on evaluated thickness of ^{26}Mg. The effect of in-beam oxygen depletion was observed in the mylar target. By analysis of yield-to-charge dependence of the first mylar exposition, a correction to the yield at zero-charge was made. The yields of reactions on ^{12}C and ^{26}Mg were charge-stable. More severe problem, was the target inhomogeneity observed in the magnesium target thickness measurement (fig. 1a). Based on this observation, RBS (Rutherford back-scattering) scanning of this target was performed at Tandetron facility of NPI CAS in Rez (CANAM infrastructure). The results of this microbeam scanning confirmed the significant inhomogeneity of magnesium layer in the target (fig. 1b). This fact makes impossible the direct target thickness measurement with alpha particle source. The above mentioned in-beam data comparison for mylar and carbon targets was used for the evaluation of the "effective"

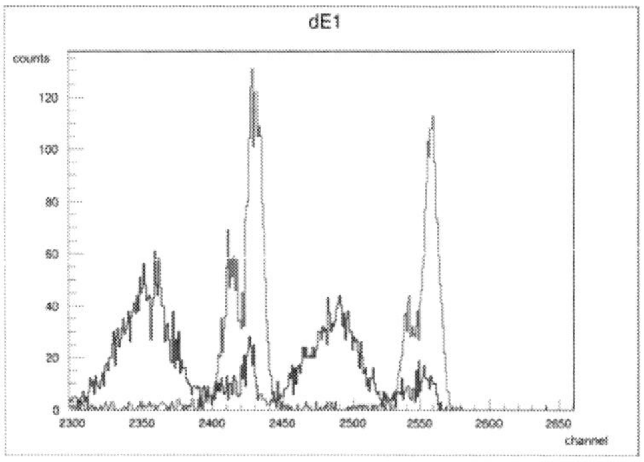

(a)

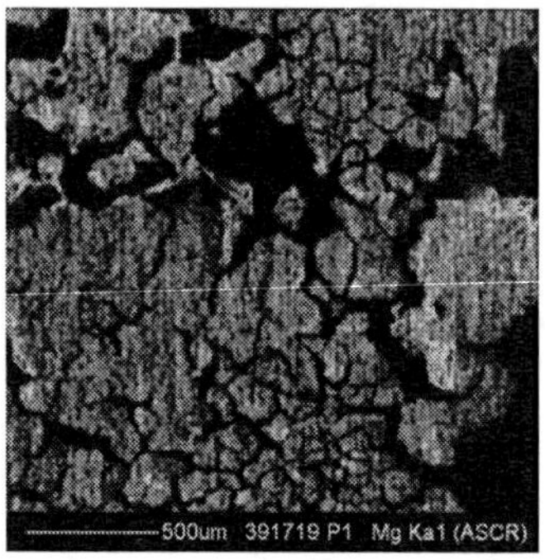

(b)

Figure 1: Analysis of magnesium layer thickness in target: (a) target thickness measurement with ^{244}Cm+^{238}Pu alpha particle source; (b) RBS analysis of magnesium layer in the target.

thickness in the magnesium target beam spot. The assumption of the chemical form of pure MgO in the target was checked, due to hygroscopic properties of this material. The beam-monitor data did not show any trend of oxygen yields in the magnesium target with the beam charge collected. There is a need for the new, good quality target. The best solution could be pure metallic form, prepared by rolling to 100-200 μm thick foils. However, preparation of this type of target would be challenging even in top laboratories. The decision was made to prepare MgO sputtering targets on carbon backing, in cooperation with Istituto Nazionale di Fisica Nucleare, Laboratori Nazionali del Sud in Catania. A set of perfect quality targets was manufactured, together with pure carbon backing targets. This method do not simplify the complicated spectra obtained from MgO+C target, but there are homogeneous, quality targets with samples of carbon backing for subtraction. A methodology of Mg target irradiation, together with mylar foil and carbon backing at the same angle, allows to quantitatively characterize the stoichiometry in the Mg target and to evaluate cross sections from the acquired spectra for the forthcoming measurements of ^{26}Mg(d,p)^{27}Mg reaction without doubts, as well as to perform a control run on (^{3}He,d) reaction to normalize the previous data from the discussed target.

5 Summary

Angular distributions of two nuclear reactions ^{26}Mg(^{3}He,d)^{27}Al and ^{26}Mg(d,p)^{27}Mg were measured. The target analysis showed the inhomogeneity of magnesium layer, which was confirmed by RBS micro-beam scanning. Based on reactions on ^{12}C and ^{16}O in mylar and carbon targets, the effective thickness of the magnesium layer was estimated. The new, high quality targets were manufactured in INFN LNS together with carbon backing samples, which allows to quantitatively evaluate angular distributions of the investigated reaction products as well as to confirm the results obtained with the previous target. This publication was supported by OP RDE, MEYS, Czech Republic under the project SPIRAL2-CZ, CZ.02.1.01/0.0/0.0/16-013/0001679.

References

[1] Champagne et al., Nuclear Physics **A512** (1990)

[2] Timofeyuk et al., Physical Review C **78** (2008) 054322
[3] A.M. Mukhamedzanov et al., Physical Review C **56** (1997) 1302
[4] C.A. Gagliardi et al., Physical Review C **59** (1999) 1149
[5] J.Mrazek, computer code calibNR, https://github.com/jaromrax/calibNR (-2018)

^{7}Be and ^{8}B reaction dynamics at Coulomb barrier energies

E. Strano[1,2,⋆], *M.* Mazzocco[1,2], *A.* Boiano[3], *C.* Boiano[4], *M.* La Commara[3,5], *C.* Manea[2], *C.* Parascandolo[3], *D.* Pierroutsakou[3], *C.* Signorini[6], *D.* Torresi[1,2], *H.* Yamaguchi[7], *D.* Kahl[7], *L.* Acosta[8,9], *P.* Di Meo[3], *J.P.* Fernandez-Garcia[9], *T.* Glodariu[10], *J.* Grebosz[11], *A.* Guglielmetti[4,12], *N.* Imai[7,13], *Y.* Hirayama[13], *H.* Ishiyama[13], *N.* Iwasa[14], *S.C.* Jeong[13,15], *H.M.* Jia[16], *N.* Keeley[17], *Y.H.* Kim[13], *S.* Kimura[13], *S.* Kubono[7,18], *J.A.* Lay[1,2], *C.J.* Lin[16], *G.* Marquinez-Duran[8], *I.* Martel[8], *H.* Miyatake[13], *M.* Mukai[13], *T.* Nakao[19], *M.* Nicoletto[2], *A.* Pakou[20], *K.* Rusek[21], *Y.* Sakaguchi[7], *A.M.* Sanchez-Benitez[8], *T.* Sava[10], *O.* Sgouros[20], *C.* Stefanini[1,2], *F.* Soramel[1,2], *V.* Soukeras[20], *E.* Stiliaris[22], *L.* Stroe[10], *T.* Teranishi[23], *N.* Toniolo[6], *Y.* Wakabayashi[18], *Y.X.* Watanabe[13], *L.* Yang[16], and *Y.Y.* Yang[24]

[1]*Dipartimento di Fisica e Astronomia, Università di Padova, Padova, Italy*
[2]*INFN-Sezione di Padova, Padova, Italy*
[3]*INFN-Sezione di Napoli, Napoli, Italy*
[4]*INFN-Sezione di Milano, Milano, Italy*
[5]*Dipartimento di Fisica, Università di Napoli "Federico II", Napoli, Italy*
[6]*INFN-Laboratori Nazionali di Legnaro (LNL), Legnaro (PD), Italy*
[7]*CNS - The University of Tokyo, RIKEN campus, Wako, Saitama, Japan*
[8]*Departamento de Fisica Aplicada, Universidad de Huelva, Huelva, Spain*
[9]*INFN-Sezione di Catania, Catania, Italy*
[10]*NIPNE, Magurele, Romania*
[11]*IFJ PAN, Krakow, Poland*
[12]*Dipartimento di Fisica, Università di Milano, Milano, Italy*
[13]*KEK, Tsukuba, Ibaraki, Japan*
[14]*Department of Physics, Sendai, Miyagi, Japan*
[15]*Institute for Basic Science, Daejeon, Korea*
[16]*China Institute of Atomic Energy, Beijing, China*
[17]*National Centre for Nuclear Research, Otwock, Poland*
[18]*RIKEN Nishina Center, Wako, Saitama, Japan*
[19]*Advanced Science Research Center, JAEA, Tokai, Ibaraki, Japan*
[20]*Department of Physics, University of Ioannina, Ioannina, Greece*
[21]*Heavy Ion Laboratory, University of Warsaw, Warsaw, Poland*
[22]*Department of Physics, University of Athens, Athens, Greece*
[23]*Department of Physics, Kyushu University, Hakozaki, Fukuoka, Japan*
[24]*Institute of Modern Physics, Chinese Academy of Sciences, Lanzhou, China*

Abstract. We investigated the reaction dynamics induced by the ^{7}Be,^{8}B+^{208}Pb collisions at energies around the Coulomb barrier. Charged particles originated by both the collisions were detected by means of 6 ΔE-E$_{res}$ telescopes of a newly developed detector array. Experimental data were analysed within the framework of the Optical Model and the total reaction cross-sections were compared together and with the ^{6,7}Li+^{208}Pb colli-

⋆e-mail: emanuele.strano@unipd.it

sion data. According to the preliminary results, ^{7}Be nucleus reactivity is rather similar to the ^{7}Li one whereas the ^{8}B+^{208}Pb total reaction cross section appears to be much larger than those measured for reactions induced by the other weakly-bound projectiles on the same target.

1 Introduction

The study of ^{7}Be and ^{8}B dynamical behavior is relevant for the understanding of several processes in the universe [1]: ^{7}Be is at the top side of the reaction network which describes the primordial nucleosynthesis in the framework of the standard Big Bang, being involved in the production of the primordial lithium; jointly with ^{8}B, is a main actor in the proton-proton chains of hydrogen burning in the main-sequence stars like our sun; in particular, ^{7}Be is at the vertex of two reaction chains, being the product of the ^{3}He+^{4}He fusion; beta decays of ^{7}Be and ^{8}B are responsible of the neutrino flux from the Sun, studied for the understanding of neutrino oscillations; finally, ^{7}Be is involved in additional reaction networks which describe the nucleosynthesis model of the inomogeneus big bang and some supernovae.

^{7}Be and ^{8}B are exotic nuclei which are also intriguing for the peculiar structures they exhibit: ^{8}B is characterized by a proton halo (which was discovered by measuring its quadrupole moment [2]); ^{7}Be, which is the core of ^{8}B, shows a cluster structure consisting of a ^{3}He-^{4}He system. These properties, jointly with a low binding energy, open the doors for a series of statical and dynamical effects which may alter the total reactivity by several order of magnitude, making of ^{7}Be,^{8}B+^{208}Pb study an effective workbench for testing nuclear models. Obviously, our comprension of the behavior of these nuclei will be extended to the astrophysical case.

For the production of dripline nuclei via (p,n), (^{3}He,n), (d,n) and (d,^{3}He) two-body reactions, the facility EXOTIC [3] was laid out in Laboratori Nazionali di Legnaro (LNL), Italy. The secondary RIB under production is separated from the scattered beam and other possible contaminant beams by means of a combination of 8 ion-optical devices and a series of slit-sets and collimators placed at suitable positions along the facility. EXOTIC is especially suitable for the production of RIBs by means of (p,n) reactions, since the larger projectile-to-target ratio ensures an additional forward focusing of the reaction products. So far we produced several beams like ^{17}F, ^{8}B, ^{7}Be (rate ~250 kHz), ^{15}O, ^{8}Li and isotopes of carbon.

At the far end of the EXOTIC beam line is usually placed the EXPADES detector array [4]. It is designed to be compact, with a high granularity and a good covered solid angle. It consists of 8 silicon telescopes arranged in a cylindrical configuration. In each telescope there are a 40 μm DSSSD acting as ΔE stage and a 300 μm DSSSD that measures the particle residual energy. Both the DSSSD are square detectors 64 mm wide and have 32 stripes per side. A set of 8 ionization chambers can be used as alternative ΔE stage.

2 ^{7}Be+^{208}Pb,^{58}Ni collisions

In the covered angular ranges [55°,84°], [96°,125°] and [138°,165°] in the laboratory frame, we measured the ^{7}Be+^{208}Pb quasi-elastic scattering since the excitation of both the projectile and the target cannot be disentangled by watching only at energy spectra. At the beginning, we extracted the angular distributions at three energies: 37.4 MeV, 40.5 MeV and 42.2 MeV. A preliminary optical model best-fit, carried out employing the code FRESCO [5], allowed the extraction of three points for the total reaction cross-section where we observed a very similar behavior to the ^{7}Li case [6], which is its mirror nucleus.

At a later time we concentrated on the ^{3}He and ^{4}He production, since they may came from several direct processes: one neutron stripping/pickup, the stripping of one of the clusters in the projectile, the exclusive breakup which is the inverse reaction of the ^{7}Be synthesis in the proton-proton chains. We clearly observed that ^{4}He production is four to five times larger than ^{3}He and identified several processes over about $3.5 \cdot 10^6$ detected events: 15 neutron pickup events and 17 ^{4}He-p coincidences. We observed only 19 exclusive break-up events, suggesting that probably the exclusive breakup is not dominant in this collision. The exclusive breakup counts are important because this process can be exploited for the extraction of the astrophysical reaction rate from the Coulomb dissociation of the projectile [7].

The ^{4}He overproduction with respect to the ^{3}He one was already observed in the case of the ^{58}Ni target. In this case we detected only few of the processes involved in the production of ^{3}He and ^{4}He and the undetected processes were evaluated using Monte Carlo simulations. In particular the exclusive breakup was evaluated using Coupled-Channel calculations. A more accurate investigation is hoped for this collision with better experimental conditions.

3 ^{8}B+^{208}Pb collision

For the ^{8}B+^{208}Pb collision we moved EXPADES to the C-RIB facility at Riken, Wako-shi, Japan where the incident beam was a "Cocktail beam" containing also ^{7}Be and ^{3}He contaminants (^{8}B rate ~10 kHz purity ~25%). However the experimental apparatus (and the C-RIB monitoring devices) provided several information that we used to separate: firstly the ^{8}B beam from its contaminants and secondly the reaction products each other. In particular we not only used the ΔE-Er matrix, which showed a very good charge separation, but also the time of flight-ΔE matrix. Also in this case we firstly concentrate only on the elastic events spot, extracting an angular distribution and a total reaction cross section. In Fig. 1 we show the very preliminary results for ^{8}B and our results for ^{7}Be. An interesting preliminary evidence is a factor three super-reactivity of ^{8}B (of about 1 b) with respect to the ^{7}Be case.

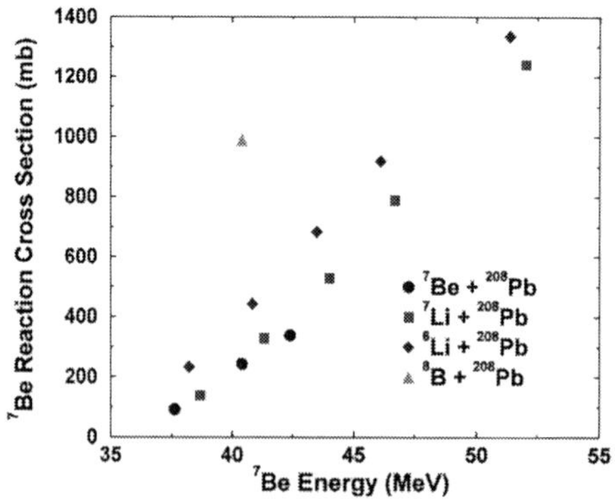

Figure 1. Very preliminary reaction cross sections for ^{6,7}Li,^{7}Be,^{8}B+^{208}Pb. Data were normalized and plotted according to Ref. [8]. Uncertainty is about 15% for ^{8}B+^{208}Pb and 10% for ^{7}Be+^{208}Pb, ^{6,7}Li+^{208}Pb tabulated without error as reference.

4 Conclusion

In this article we showed some results about two light exotic nuclei which are involved in several astrophysical processes. We studied both nuclei performing their collision on a ^{208}Pb target and employing an high granularity detection array. We observed that in the case of ^{7}Be dynamics is not dominated by the breakup process, observing 19 coincidences and, since this process is useful for the determination of the astrophysical reaction rates, more efforts are required in measuring this process. Monte Carlo simulations are presently being performed in order to determinate the cross-section of each process. In the case of ^{8}B, preliminary results show a factor three larger reactivity with respect to the ^{7}Be case. Analysis is still going on and interesting results are expected.

References

[1] H. Oberhummer et al., Few-Body Syst. Suppl. **8**, 317 (1995), nucl-th/9510031
[2] T. Minamisono et al., Phys. Rev. Lett. **69**, 2058 (1992)
[3] F. Farinon et al., Nucl. Inst. Meth. B **266**, 4097 (2008)
[4] D. Pierroutsakou et al., Nucl. Instr. Meth. Phys. Res. A **834**, 46 (2016)
[5] I.J. Thompson, Computer Physics Reports **7**, 167 (1988)
[6] N. Keeley et al., Nuclear Physics A **571**, 326 (1994)
[7] S. Kubono, Nuclear Physics A **693**, 221 (2001), radioactive Nuclear Beams
[8] P.R.S. Gomes et al., Phys. Rev. C **71**, 017601 (2005)

AUTHOR INDEX

Abe K. 152, 173
Aberle O. 144
Acosta L. 28, 180
Adachi S. 152
Adsley P. 162
Agodi C. 28
Alellara A. 162
Andrzejewski J. 139, 144
Anzalone A. 149, 156
Aoi N. 152
Aoki Wako 1
Asai M. 152
Assié M. 162
Assunção M. 162
Audouin L. 144
Bacak M. 144
Balabanski D. 149
Balabanski D.L. 156
Balibrea J. 144
Barbagallo M. 139, 144
Bastin B. 162
Beaumel D. 162
Becvár F. 144
Beliuskina O. 173
Béroff K. 162
Berthoumieux E. 144
Bertulani C.A. 7
Billowes J. 144
Bishop Shawn 53
Boiano A. 180
Boiano C. 180
Bosnar D. 144
Broggini Carlo 16
Brown A. 144
Burjan V. 102, 134, 176
Caamaño M. 144
Calviani M. 144
Calviño F. 144
Cano-Ott D. 144
Cappuzzello F. 28
Carbone D. 28
Cardella R. 144
Casanovas A. 144
Cavallaro M. 28
Cerutti F. 144
Cha S. M. 173
Chae K. Y. 173
Chen A. A. 173
Chen Y. H. 144
Cherubini S. 28, 130, 134, 169
Cherubini Silvio 53, 114
Chesnevskaya S. 149, 156
Chiaveri E. 144
Chillery Thomas 126
Chintak Shubh 7
Coc A. 162
Coc Alain 53
Colonna N. 139, 144
Cortés G. 144
Cortés-Giraldo M. A 144
Cosentino L. 139, 144
Costa M. 156
Cristallo Sergio 22
Crucillà V. 156
Crucillá W. 149
Cvetinovic A. 28, 130, 169
D'Agata G. 28, 102, 134, 162, 169, 176
D'Agata Giuseppe 114
Damone L. A. 144
De Oliveira F. 162
de Oliveira Santos Francois 36
de Séréville N. 162
de Séréville Nicolas 53
Delafosse C. 162
Dell'Aquila D. 28, 169
Di Meo P. 180
Di Pietro A. 134, 169
Diakaki M. 144
Dietz M. 144
Dinh Trong T. 152
Domingo-Pardo C. 144
Dressler R. 144
Dupont E. 144
Durán I. 144
El Eid Mounib 159
El Eid Mounib F. 46
Famiano Michael 166
Fernández-Dominguez B. 144
Fernandez-Garcia J.P. 180
Ferrari A. 144
Ferreira P. 144
Figuera P. 169
Filipescu D. 149
Finocchiaro P. 139, 144
Flavigny F. 162
Fox S. P. 162
Francalanza L. 28
Fukuda M. 152
Furman V. 144
Gabriele Rapisarda Giuseppe 53
Gangopadhyay Mayukh 79
García A. R. 144
Garg R. 144
Gawlik A. 144
Ge Z. 173
Georgiadou A. 162
Gey G. 152
Gianluca Pizzone Rosario 53
Gilardoni S. 144
Giovanni Luca Guardo 114

AUTHOR INDEX

Giuseppe Gabriele Rapisarda 114
Glagolev V. 102, 176
Glodariu T. 144, 180
Göbel K. 144
Goncalves I. F. 144
González-Romero E. 144
Gottardo A. 162
Grassi L. 162
Grebosz J. 180
Griesmayer E. 144
Guardo G.L. 28, 130, 134, 149, 156
Guardo L. 169
Guerrero C. 144
Guglielmetti A. 180
Guillot J. 162
Guimaraes V. 162
Gulino M. 28, 130, 134, 149, 156, 169
Gulino Marisa 53, 114
Gunsing F. 144
Hammache F. 162
Hammache Fairouz 53, 59
Harada H. 144
Hashimoto T. 152
Hayakawa S. 120, 173
Hayakawa Seya 53
Heinitz S. 144
Heyse J. 144
Hirayama Y. 180
Hubbard N. 162
Id Barkach T. 162
Ideguchi E. 152
Imai N. 173, 180
Indelicato I. 28, 134, 169
Indelicato Iolanda 53, 114
Inoue A. 152
Isaak J. 152
Ishiyama H. 180
Iwasa N. 173, 180
Iwasa Naohito 53
Jenkins D. G. 144
Jeong S.C. 180
Jericha E. 144
Jia H.M. 180
Kadi Y. 144
Kahl D. 120, 144, 173, 180
Kajino Toshitaka 79, 166
Kalamara A. 144
Käppeler F. 144
Kato Seigo 53
Kavrigin P. 144
Kawag M. S. 173
Keeley N. 180
Kiener J. 162
Kim A. 173
Kim D. H. 173
Kim M. J. 173
Kim Y.H. 180
Kimura A. 144
Kimura S. 180
Kiss Gabor 53
Kivel N. 144
Kobayashi N. 152
Kokkoris M. 144
Komatsubara Tetsuro 53
Kres I. 169
Kroha V. 102, 176
Krticka M. 144
Kubono S. 173, 180
Kubono Shigeru 53, 73
Kurtulgil D. 144
Kusakabe Motohiko 79
La Cognata M. 28, 102, 130, 134, 149, 156, 169, 176
La Cognata Marco 53, 114
La Commara M. 134, 169, 180
Laird A. M. 162
Lamia L. 28, 102, 130, 134, 169, 176
Lamia Livio 53, 114
Lattuada D. 149, 156, 169
Lattuada M. 134, 169
Lay J.A. 180
Leal-Cidoncha E. 144
Lederer-Woods C. 144
Leeb H. 144
Lerendegui-Marco J. 144
Liang J. 173
Lin C.J. 180
Lo Meo S. 144
Lombardo I. 28, 169
Lonsdale S. J 144
Lucia Anna Damone 139
MacCormick M. 162
Macina D. 144
Maeda Y. 152
Makii H. 152
Makki Tahani 159
Manea C. 180
Marganiec J. 144
Maria Letizia Sergi 114
Marquinez-Duran G. 180
Marte I. 180
Martínez T. 144
Masi A. 144
Massimi C. 144
Mastinu P. 144
Mastromarco M. 139, 144
Matea I. 162
Matei C. 149, 156
Mathews Grant 79
Matsuta K. 152
Maugeri E. A. 144

AUTHOR INDEX

Mazzocco M. 134, 169, 180
Mazzocco Marco 89
Mazzone A. 144
Mendoza E. 144
Mengoni A. 144
Meyer A. 162
Mihara M. 152
Milazzo P. M. 144
Milin M. 134
Mingrone F. 144
Miura M. 152
Miyatake H. 180
Möller Peter 94
Moon J. Y. 173
Mori Kanji 166
Mrazek J. 102, 134
Mrázek J. 176
Mukai M. 180
Mukhamedzhanov A.M. 7, 102
Musumarra A. 144
Nakao T. 180
Negret A. 144
Nguyen Binh Dam 53
Nicoletto M. 180
Nishimura S. 173
Nolte R. 144
Oka S. 173
Olivier L. 162
Oprea A. 144
Ordine A. 28
Oulebsir N. 162
Pakou A. 180
Palmerini S. 134
Parascandolo C. 134, 180
Parascandolo T. 169
Parikh A. 162
Park S. Y. 173
Patronis N. 144
Pavlik A. 144
Perkowski J. 144
Perrot L. 162
Perrotta S. S. 169
Pierroutsakou D. 134, 180
Pierroutsakou Dimitra 108
Piskor Š. 102, 176
Pizzone G.R. 102
Pizzone R. G. 134, 149, 169
Pizzone R.G. 28, 130, 156, 176
Porras I. 144
Portail C. 162
Praena J. 144
Psaltis A. 173
Puglia S.M.R. 28, 130
Quesada J. M. 144
Radeck D. 144
Rapisard G.G. 176
Rapisarda G. 102
Rapisarda G.G. 28, 130, 134, 149, 169
Rauscher T. 144
Reifarth R. 144
Riley J. 162
Roeder Brian 53
Romano S. 28, 102, 130, 134, 149, 156, 169, 176
Romano Stefano 114
Rosario Gianluca Pizzone 114
Rubbia C. 144
Rusek K. 180
Ryan J. A. 144
Sabateé-Gilarte M. 144
Sakaguchi Y. 180
Sánchez Benítez Á. M. 162
Sanchez-Benitez A.M. 180
Santagati G. 28
Sasankan Nishanth 79
Sava T. 180
Saxena A. 144
Schillebeeckx P. 144
Schumann D. 144
Sedyshev P. 144
Sergi M.L. 130, 134
Sgouros O. 180
Shima T. 152
Shimizu H. 120, 152, 173
Signorini C. 180
Sivácek I. 102, 176
Smith A. G 144
Soic N. 134
Soramel F. 180
Sosnin N. V 144
Soukeras V. 180
Spadaccini G. 28
Spartá R. 28, 102, 130, 134, 169, 176
Spartà Roberta 53, 114
Spitaleri C. 28, 102, 130, 134, 149, 156, 169, 176
Spitaleri Claudio 53, 114
Stamatopoulos A. 144
Stefan I. 162
Stefanini C. 180
Stiliaris E. 180
Strano E. 180
Stroe L. 180
Taffara A. 149
Tagliente G. 144
Tain J. L 144
Tamii A. 152
Tang R. 152
Tarifeño-Saldivia A. 144
Tassan-Got L. 144
Tatischeff V. 162
Teranishi T. 173, 180

AUTHOR INDEX

Teranishi Takashi 53
Toniolo N. 180
Torresi D. 180
Trache Livius 53
Tribble R. 102
Tribble Robert..53
Trippella O. 134, 169
Tumino A. 28, 102, 130, 134, 149, 156, 169, 176
Tumino Aurora 114
Ueno Y. 173
Valenta S. 144
Vannini G. 144
Variale V. 144
Vaz P. 144
Ventura A. 144
Verde G. 28
Vigilante M. 28
Vlachoudis V. 144
Vlastou R. 144
Wakabayashi Y. 180
Wakabayashi Yasuo...53
Wallner A. 144
Warren S. 144
Watanabe Y.X. 180
Weiss C. 144
Woods P. J 144
Wright T. 144
Xu Y 156
Xu Y. 149
Yamaguchi H. 120, 152, 173, 180
Yamaguchi Hidetoshi..53
Yang L. 120, 152, 173, 180
Yang Y.Y. 180
Žugec P. 144